The Real Number System

John M. H. Olmsted

Dover Publications, Inc.
Mineola, New York

Bibliographical Note

This Dover edition, first published in 2018, is an unabridged republication of the work originally published in the Appleton-Century Monographs in Mathematics series by Appleton-Century-Crofts, a division of Meredith Publishing Company, New York, in 1962.

Library of Congress Cataloging-in-Publication Data

Names: Olmsted, John M. H. (John Meigs Hubbell), 1911–1997, author.
Title: The real number system / John M. H. Olmsted.
Description: Dover edition. | Mineola, New York : Dover Publications, 2018. |
  Originally published: New York, N.Y. : Appleton-Century-Crofts, 1962. |
  Includes index.
Identifiers: LCCN 2018009572| ISBN 9780486827643 | ISBN 048682764X
Subjects: LCSH: Numbers, Real. | Numbers, Real—Problems, exercises, etc.
Classification: LCC QA255 .O43 2018 | DDC 512.7/86—dc23
LC record available at https://lccn.loc.gov/2018009572

Manufactured in the United States by LSC Communications
82764X02   2019
www.doverpublications.com

PREFACE

++

THE NUMBERS THAT CONSTITUTE what is known as the *real number system* are basic to all of mathematics and hence to all of science. These are the numbers of the everyday world as well, some being used by the nursery child in counting piles of blocks, some by the carpenter in measuring lengths and areas, and some by the banker in reckoning profits, losses, and percentages. Properties of the real numbers permeate our culture and are used almost universally, to a varying extent and with a varying degree of explicit awareness on the part of the user. Questions naturally formulate themselves. "Why are these properties true?" "Where do they come from?" "What other properties do the real numbers have?" "Just what is a real number anyway?"

The present volume tells about real numbers by investigating the kind of structure that is formed by the *totality* of all real numbers. The real number system is described categorically in the book by a set of axioms, stated briefly: *the real number system is a complete ordered field.* These axioms are presented and studied step by step, with numerous examples used to illustrate the new concepts as they are introduced. Examples of ordered fields that are not complete, fields that are not ordered, and systems that are similar to fields but are not fields are all examined with a view to an increased understanding of the real number system itself. The discussion is extensive and multiple-leveled, and is designed for more than one particular audience. As a monograph the book attempts to tell a fairly full story, including much about induction and going all the way to the laws of exponents for arbitrary positive bases and real exponents. An appendix shows how the property of completeness can be described in a large number of equivalent fashions.

A continuous effort is made to communicate to educated laymen as much of the spirit and content of the particular axiomatic system to which the book is dedicated as is consistent with the readers' mathematical backgrounds and tastes. The discussion and proofs throughout the book are kept on as elementary a level as possible. Logical ideas are discussed in context as they arise. For example, the precise meaning of the implication *p implies q* and its relation to the "contrapositive" *not q implies not p* are explained carefully in the first chapter with the aid of examples. The general reader should find things going pretty smoothly for at least the first two chapters. Although very little specific background is required in the main part of the text, the reader who has had little contact with mathematical ideas and abstractions will probably begin to encounter occasional rough spots beginning in Chapter 3 and becoming more frequent as he progresses. With effort and concentration, however, anyone with a year or so of high school mathematics and a healthy ambition should be able to work his way steadily through to the Appendix. (The Appendix is another matter, demanding a considerable background of college mathematical analysis.)

The book may be used as a text on any of several levels. The first two chapters are at an Elementary level (high school seniors and up). With the exception of the last part of Chapter 7 (§§711–714), the most advanced material to be encountered in the first eight chapters is at an Intermediate level (advanced high school seniors, college freshmen and up). Chapters 9 and 12 should be accessible to college sophomores. The remaining parts of the main text (§§711–714 and Chapters 10 and 11) are Advanced level (college juniors and up). The entire text of twelve chapters is thus suitable for college courses at different levels, from the freshmen year through the senior or first graduate year. The Appendix is at the Graduate (or advanced undergraduate) level. The entire book should be quite appropriate for a beginning graduate student, who could absorb the first few chapters in short order, slowing down as the material increases in difficulty.

Most of the revised curricula that have been introduced recently into high schools and colleges place heavy emphasis on at least some aspects of the real number system. As a general principle it is important that anybody who is planning on teaching a particular body of material be familiar with relevant subject matter well beyond the extent to which he expects to teach it. It is in part to future teachers of high school mathematics that this book is addressed. High school teachers returning to the classroom as students—in summer schools, institutes, workshops, and the like—may also profit from a study of the book. The author has kept in mind recommendations of the Committee on the Undergraduate Program in Mathematics of the Mathematical Association of America. Such items as the following are introduced as needed, explained carefully with illustrations, and developed and applied toward an integrated and unified goal with a modern flavor: sets, Cartesian products, relations, functions, the Euclidean algorithm, greatest common divisors, equivalence relations, congruences, polynomials, rational functions, intervals, absolute values, inequalities, set-builder notation, powers of numbers, logarithms, decimal expansions, and representations of numbers with bases other than ten.

A high level of precision and completeness characterizes the text. Proofs are given in full and rigorous detail, except for those that are parallel to proofs already given. In many such instances proofs are requested in the exercises. In general, the exercises are ample (there are 371 of them) and well-graded, permitting the reader to test and develop his knowledge. Answers are given in the back of the book. Illustrative examples are generously provided.

A system of starring ($\star$) permits considerable flexibility in the selection of material. Whereas in some books starring is used to indicate sections or exercises that are more difficult or advanced than others, this is not the case in the present volume, where the role of starring is simply that of route-marking. A star attached to a portion of this book has the following single significance: *that particular starred portion is not required in any unstarred portion.* In other words, prerequisite to any unstarred section are the preceding unstarred sections, while prerequisite to any starred section are *all* preceding sections. Thus, the following two routes (and combinations thereof) are possible: (1) straight through the book to whatever extent time, preparation, and ability permit; (2) straight through the unstarred sections of the book to whatever extent time, preparation, and ability permit. The second route leads more directly to the completeness property, and hence to the full set of axioms for the real number system. Each of these two routes is logically self-contained if followed without gaps.

As a kind of recapitulation of some of the comments made above, the following seven courses typify what is possible with this book as text, where Route 1 places emphasis on congruences, finite fields, and polynomials, and Route 2 places emphasis on a direct path to the completeness axiom, which is of basic importance in calculus and advanced mathematical analysis:

TABLE OF POSSIBLE COURSES

Route 1. Starred and Unstarred Sections
Route 2. Unstarred Sections

Course	Academic Level of Students	Route	Portions of Book
I	High School Students	2	Chapters 1 and 2
II	Advanced High School Students or College Freshmen	1	Chapters 1–8, omitting §§ 711–714
III	College Sophomores	1	Chapters 1–9, 12, omitting §§ 711–714
IV	College Sophomores	2	Chapters 1–5, 8, 9, 12, omitting §§ 407–409
V	College Juniors and up, or Institutes for High School Teachers	1	All chapters except Appendix (as time permits)
VI	College Juniors and up, or Institutes for High School Teachers	2	Chapters 1–5, 8–12, omitting §§ 407–409 (as time permits)
VII	Graduates and Advanced Undergraduates	1	Chapters 1–8 (review), 9–11, 12 optional Appendix emphasized

A few words regarding notation should be given. The equal sign $=$ is used for equations, both conditional and identical, and the triple bar $\equiv$ is reserved for definitions. For simplicity, if the meaning is clear from the context, neither symbol is restricted to the indicative mood as in "$(a + b)^2 = a^2 + 2ab + b^2$," or "where $f(x) \equiv x^2 + 5$." Examples of subjunctive uses are "let $x = n$," and "let $m \equiv 1$," which would be read "let x be equal to n," and "let m be defined to be 1," respectively. A similar freedom is granted the inequality symbols. For instance, the symbol $>$ in the following constructions "if $x > 0$, then $\cdots$," "let $x > 0$," and "let $x > 0$ be given," could be translated "is greater than," "be greater than," and "greater than," respectively.

The author wishes to express his appreciation of the aid and suggestions given by Professor R. W. Brink of the University of Minnesota in the preparation of the manuscript. He is also indebted to many others for their helpful comments.

J.M.H.O.

Carbondale, Illinois

CONTENTS

++

ix

Chapter 4

COMPOSITE AND PRIME NUMBERS

Chapter 5

INTEGERS AND RATIONAL NUMBERS

★Chapter 6

CONGRUENCES AND FINITE FIELDS

★Chapter 7

POLYNOMIALS AND RATIONAL FUNCTIONS

Chapter 8
INTERVALS AND ABSOLUTE VALUE

Chapter 9
THE AXIOM OF COMPLETENESS

Chapter 10
ROOTS AND RATIONAL EXPONENTS

Chapter 11
EXPONENTS AND LOGARITHMS

Chapter 12

DECIMAL EXPANSIONS

APPENDIX

1

Fields

101. INTRODUCTION

Any mathematical discipline must rest ultimately on certain undefined objects and unproved statements—in short, on an *axiomatic system*. The basis upon which that part of mathematics known as analysis rests is the system of *real numbers*, numbers like 1, 2, $\sqrt{2}$, $\sqrt[3]{5}$, π, and their negatives, used in quantitative measurement. There are two principal ways of studying the real number system. One method is to start with a more primitive system—such as the set of *natural numbers (positive integers)* 1, 2, 3, $\cdots$— and from this system, by a sequence of definitions and theorems, to build or construct the more elaborate system of real numbers.† The question of *existence* of the real number system, then, is replaced by the question of existence of the natural numbers and acceptance of the laws of logic involved in the constructive process. The second method is to give a formal description of the real number system (assumed to exist) by means of a *fundamental set* of properties from which all other properties can be derived. It is this latter *descriptive* approach that has been chosen for this book. The fundamental set of properties, known as the **axioms** of the real numbers, will be presented in parts and explored gradually. It is our ultimate purpose to learn precisely what is meant when it is said that "the real number system is a complete ordered field." We start with a field. (Ordering is discussed in Chapter 2, and completeness in Chapter 9.)

† The Italian mathematician G Peano (1858–1932) presented in 1889 a set of five axioms for the natural numbers. For a detailed discussion of the development of the real numbers from the Peano axioms, see F. Landau, *Foundations of Analysis* (New York, Chelsea Publishing Co., 1951). Other books that treat the real number system are G. Birkhoff and S. MacLane, *A Survey of Modern Algebra* (New York, The Macmillan Company, 1953); N. T. Hamilton and J. Landin, *Set Theory, the Structure of Arithmetic* (Boston, Allyn and Bacon, Inc., 1961); R. B. Kershner and L. R. Wilcox, *The Anatomy of Mathematics* (New York, Ronald Press Co., 1950); H. Levi, *Elements of Algebra* (New York, Chelsea Publishing Co., 1953); M. H. Maria, *The Structure of Arithmetic and Algebra* (New York, John Wiley & Sons, Inc., 1958); I. Niven, *Numbers: Rational and Irrational* (New York, Random House, 1961); J. B. Roberts, *The Real Number System in an Algebraic Setting* (San Francisco, W. H. Freeman and Co., 1962); M. H. Stone, *The Theory of Real Functions* (Cambridge, Mass., Harvard University Mathematics Department, 1940); and H. A. Thurston, *The Number-System* (New York, Interscience Publishers Inc., 1956).

102. AXIOMS OF A FIELD

The basic operations of the number system are *addition* and *multiplication*. As shown below, *subtraction* and *division* can be defined in terms of these two. The axioms of the basic operations are divided into three categories: *addition*, *multiplication*, and *addition and multiplication*.

The following sets of axioms, I, II, and III, define what is known as a **field**. That is, any set of objects (herein referred to simply as **numbers**) with two operations satisfying these axioms is *by definition* a field. The symbol $\mathcal{R}$ will be used to designate the real number system or, for purposes of this chapter, any field.

I. Addition

(*i*) *Any two numbers, x and y, have a unique sum, $x + y$.*

(*ii*) *The **associative law** holds. That is, if x, y, and z are any numbers,*

$$x + (y + z) = (x + y) + z.$$

(*iii*) *There exists a number 0 such that for any number x, $x + 0 = x$.*

(*iv*) *Corresponding to any number x there exists a number $-x$ such that*

$$x + (-x) = 0.$$

(*v*) *The **commutative law** holds. That is, if x and y are any numbers,*

$$x + y = y + x.$$

The number 0 of axiom (*iii*) is called **zero**.

The number $-x$ of axiom (*iv*) is called the **negative** of the number x. The **difference** between x and y is defined:

$$x - y \equiv x + (-y).$$

The resulting operation is called **subtraction**.

Some of the properties that can be derived from Axioms I alone are given in Examples 1 and 2, § 103, and Exercises 1–5, § 104.

II. Multiplication

(*i*) *Any two numbers, x and y, have a unique product, xy, also denoted $x \cdot y$.*

(*ii*) *The **associative law** holds. That is, if x, y, and z are any numbers,*

$$x(yz) = (xy)z.$$

(*iii*) *There exists a number $1 \neq 0$ such that for any number x, $x \cdot 1 = x$.*

(*iv*) *Corresponding to any number $x \neq 0$ there exists a number x^{-1} such that $x \cdot x^{-1} = 1$.*

(*v*) *The **commutative law** holds. That is, if x and y are any numbers,*

$$xy = yx.$$

The number 1 of axiom (*iii*) is called **one** or **unity**.

The number x^{-1} of axiom (*iv*) is called the **reciprocal** of the number x. The **quotient** of x and y ($y \neq 0$) is defined:

$$\frac{x}{y} = x/y \equiv x \cdot y^{-1}.$$

The resulting operation is called **division**.

Some of the properties that can be derived from Axioms II alone are given in Exercises 6–9, § 104.

III. Addition and Multiplication

*The **distributive law**† holds. That is, if x, y, and z are any numbers,*

$$x(y + z) = xy + xz.$$

The distributive law, together with Axioms I and II, yields further familiar relations, some of which are given in Examples 3–7, § 103, and Exercises 10–29, § 104. The reader should notice in particular Exercises 15 and 16, § 104, in which the standard procedures for multiplying and adding fractions are established.

NOTE. The operations of addition and multiplication are called **binary operations** because each is applied to a *pair* of numbers.

103. SOME SIMPLE CONSEQUENCES

In the examples given below, we shall demonstrate how some of the most familiar properties of numbers are direct consequences of the fact that *the real number system is a field*. In the Exercises of § 104 the reader has the opportunity to work out proofs of several more statements. Many of these proofs involve techniques illustrated in the following examples. Scattered hints are provided.

In the Exercises of § 104, as well as in subsequent sets of exercises, *all statements will be considered available for later use*, even though their proofs are the responsibility of the reader.

Example 1. Prove the uniqueness of zero. That is, prove that there is only one number having the property of the number 0 of Axiom I(*iii*).

Solution. Assume that the numbers 0 and 0′ both have the property specified. Then, since $x + 0 = x$ for every number x, this equality is true in particular when $x = 0'$:

(1) $0' + 0 = 0'.$

Similarly, since $x + 0' = x$ for every x,

(2) $0 + 0' = 0.$

---

† In more precise technical language: "Multiplication is distributive with respect to addition." (Cf. the Birkhoff and MacLane book cited in the footnote of § 101.)

By the commutative law for addition, the left-hand members of (1) and (2) are equal. Since each of these equations expresses an equality between two numbers, that is, states that the numbers on the two sides of that equation are the same number, we can conclude that the right-hand members of (1) and (2) are the same number. (Things equal to the same thing are equal to each other.) This is the desired conclusion.

Example 2. Prove the **law of cancellation for addition:**

$$x + y = x + z \text{ implies } y = z.$$

Solution. We start by assuming that $x + y = x + z$; in other words, we assume that $x + y$ and $x + z$ are merely *two symbols for the same number.* Let $(-x)$ be a number satisfying Axiom I(iv). Then, since $x + y$ and $x + z$ are the same number, the sum of $(-x)$ and this number can be written in either of the following two equivalent ways:

$$(3) \qquad\qquad (-x) + (x + y) = (-x) + (x + z)$$

(if the same quantity is added to equal quantities the sums are equal). By the associative law for addition, the two members of (3) can be rewritten to give the equation

$$(4) \qquad\qquad [(-x) + x] + y = [(-x) + x] + z.$$

By the commutative law the quantity in square brackets, on each side of (4), is equal to $x + (-x)$, and therefore to 0, and equation (4) becomes

$$(5) \qquad\qquad 0 + y = 0 + z.$$

Again, by the commutative law for addition,

$$(6) \qquad\qquad y + 0 = z + 0,$$

and we have the desired conclusion: $y = z$.

Example 3. Prove that for any number x:

$$(7) \qquad\qquad x \cdot 0 = 0.$$

Solution. We start with the equation

$$(8) \qquad\qquad 0 + 0 = 0,$$

and multiply the number x by the number 0 using each of its two representations in (8) (if the same quantity is multiplied by equal quantities the products are equal):

$$(9) \qquad\qquad x(0 + 0) = x \cdot 0.$$

By the distributive law:

$$(10) \qquad\qquad x \cdot 0 + x \cdot 0 = x \cdot 0.$$

On the other hand, by the defining property I(iii) of 0,

$$(11) \qquad\qquad x \cdot 0 + 0 = x \cdot 0.$$

Since the right-hand members in (10) and (11) are the same, so are their left-hand members:

$$(12) \qquad\qquad x \cdot 0 + x \cdot 0 = x \cdot 0 + 0.$$

Finally, by the law of cancellation for addition, established in Example 2, we have the desired conclusion, equation (7).

Example 4. Prove that the product of two nonzero numbers is nonzero. That is, if x is not equal to 0 and if y is not equal to 0, then xy is not equal to 0 or, expressed slightly differently, $x \neq 0$ and $y \neq 0$ together imply $xy \neq 0$.

Solution. An implication of the form *if p then q*, or *p implies q*, in mathematics means that it is impossible for p to be true and q to be false simultaneously. In the present example the letter p stands for the compound statement $x \neq 0$ and $y \neq 0$, and the letter q stands for the statement $xy \neq 0$. As is often expedient in establishing an implication, we shall show in this case that p *does* imply q by assuming that p does *not* imply q; that is, that it is possible for p to be true and q to be false at the same time, and then obtain a contradiction. To be precise, we assume that there are two numbers x and y such that $x \neq 0$ and $y \neq 0$ and $xy = 0$. By multiplying both members of this last equation by x^{-1}, and using Example 3, we have:

$$x^{-1}(xy) = x^{-1} \cdot 0 = 0.$$

On the other hand, by the associative and commutative laws for multiplication:

$$x^{-1}(xy) = (x^{-1}x)y = (xx^{-1})y = 1 \cdot y = y \cdot 1 = y.$$

Consequently $y = 0$, in contradiction to one of the hypotheses. This means that in the presence of the assumptions $x \neq 0$ and $y \neq 0$ the equation $xy = 0$ is impossible.

Example 5. Prove that if $xy = 0$, then either $x = 0$ or $y = 0$ (or both $x = 0$ and $y = 0$).

Solution. This implication is logically equivalent to that of Example 4. In order to see this, let us express the present implication in the form *A implies B*, where A is the statement $xy = 0$ and B is the statement that either $x = 0$ or $y = 0$ (or both $x = 0$ and $y = 0$). Our objective, then, is to show that it is impossible for A to be true and B to be false simultaneously or, in other words, that it is impossible for two numbers x and y to exist such that their product xy is equal to 0 and such that it is false that at least one of them is equal to 0. In still different language, we are to establish the impossibility of having two numbers x and y whose product xy is equal to 0 at the same time that *neither* one of them is equal to 0: $x \neq 0$ and $y \neq 0$. But this impossibility is precisely the one established in Example 4. The reason for this situation can be described logically as follows: The statements p of Example 4 and B of this example are negations of each other—each is the direct denial of the other, written $p = $ *not B* or $B = $ *not p*—and the statements q of Example 4 and A of this example are negations of each other, $q = $ *not A* or $A = $ *not q*. The implication of the present example can therefore be expressed in the form *not q implies not p*; it takes the form of the impossibility that *not q* is true and *not p* is false simultaneously or, in other words, of the impossibility that p is true and q is false simultaneously, and we have the implication *p implies q* of Example 4. The implication *not q implies not p* is called in logic the **contrapositive** of the implication *p implies q*, and is logically equivalent to it, for reasons outlined above. In summary, then, the statement to be established in the present example is true since it is the contrapositive of the statement already established in Example 4.

Example 6.   Prove that if $x \neq 0$, then $x^{-1} \neq 0$.

Solution.   We seek a contradiction to the assumption that there exists a number x such that $x \neq 0$ and $x^{-1} = 0$.   By assumption, then, the product xx^{-1} is equal to $x \cdot 0$ which, by Example 3, is equal to 0.   On the other hand, by definition of x^{-1}, this same product is equal to unity: $xx^{-1} = 1$.   But this means that $1 = 0$, in contradiction to Axiom II(*iii*)!   With this contradiction the implication $x \neq 0$ *implies* $x^{-1} \neq 0$ is established.

Example 7.   Prove that if $y \neq 0$, then $\dfrac{x}{y} = 0$ if and only if $x = 0$.

Solution.   The words "if and only if" indicate a two-way implication.   We prove the "if" part first by assuming $x = 0$ and proving that as a consequence $x/y = xy^{-1}$ $= 0$.   This last equality is true since, by assumption, $xy^{-1} = 0y^{-1}$, and this is equal to 0 by Example 3: $0y^{-1} = y^{-1}0 = 0$.   For the "only if" part of the proof we reverse the implication and assume that $xy^{-1} = 0$, and wish to conclude that $x = 0$. By Example 5, we know that *either* $x = 0$ or $y^{-1} = 0$, and by Example 6, $y^{-1} \neq 0$. Therefore, we are forced to the conclusion $x = 0$, as desired.

104. EXERCISES

In Exercises 1–29, prove the given statement or establish the given equation for an arbitrary field.

1. The negative of a number is unique.   *Hint:* If y has the property of $-x$ in Axiom I(*iv*), § 102, $x + y = x + (-x) = 0$.   Use the law of cancellation, given in Example 2, § 102.

2. $-0 = 0$.   *Hint:* $0 + 0 = 0 + (-0) = 0$.

3. $-(-x) = x$.   *Hint:* $(-x) + x = (-x) + [-(-x)] = 0$.

4. $0 - x = -x$.

5. $-(x + y) = -x - y$; $-(x - y) = y - x$.   *Hint:* By the uniqueness of the negative it is sufficient for the first part to prove that
$$(x + y) + [(-x) + (-y)] = 0.$$
Use the commutative and associative laws.

6. There is only one number having the property of the number 1 of Axiom II(*iii*), § 102.

7. The **law of cancellation for multiplication** holds: $xy = xz$ implies $y = z$ if $x \neq 0$.

8. The reciprocal of a number ($\neq 0$) is unique.

9. $1^{-1} = 1$.

10. Zero has no reciprocal.

11. If $x \neq 0$, $(x^{-1})^{-1} = x$.

12. $\dfrac{1}{x} = x^{-1} \, (x \neq 0)$.

13. If $x \neq 0$ and $y \neq 0$, then $(xy)^{-1} = x^{-1}y^{-1}$, or $\dfrac{1}{xy} = \dfrac{1}{x} \cdot \dfrac{1}{y}$.

14. If $b \neq 0$ and $d \neq 0$, then $\dfrac{a}{b} = \dfrac{ad}{bd}$.

Hint: $(ad)(bd)^{-1} = (ad)(d^{-1}b^{-1}) = a[(dd^{-1})b^{-1}] = ab^{-1}$.

15. If $b \neq 0$ and $d \neq 0$, then $\dfrac{a}{b} \cdot \dfrac{c}{d} = \dfrac{ac}{bd}$.

16. If $b \neq 0$ and $d \neq 0$, then $\dfrac{a}{b} + \dfrac{c}{d} = \dfrac{ad + bc}{bd}$.

Hint: $(bd)^{-1}(ad + bc) = (b^{-1}d^{-1})(ad) + (b^{-1}d^{-1})(bc)$.

17. $(-1)(-1) = 1$. *Hint:* $(-1)(1 + (-1)) = 0$. The distributive law gives $(-1) + (-1)(-1) = 0$. Add 1 to each member.

18. $(-1)x = -x$. *Hint:* Multiply each member of the equation $1 + (-1) = 0$ by x.

19. $(-x)(-y) = xy$. *Hint:* Write $-x = (-1)x$ and $-y = (-1)y$.

20. $-(xy) = (-x)y = x(-y)$.

21. $-\dfrac{x}{y} = \dfrac{-x}{y} = \dfrac{x}{-y}$ $(y \neq 0)$.

22. $x(y - z) = xy - xz$.

23. $(x - y) + (y - z) = x - z$.

24. $(a - b) - (c - d) = (a + d) - (b + c)$.

25. $(a + b)(c + d) = (ac + bd) + (ad + bc)$.

26. $(a - b)(c - d) = (ac + bd) - (ad + bc)$.

27. $a - b = c - d$ if and only if $a + d = b + c$.

28. If $x^2 = x \cdot x$, $x^2 - y^2 = (x - y)(x + y)$.

29. The general linear equation $ax + b = 0$, $a \neq 0$, has a unique solution $x = -b/a$.

30. Axioms I(*i*), (*ii*), (*iii*), and (*iv*), § 102, define a **group**. That is, any set of objects with one operation (in this case addition) satisfying these axioms is *by definition* a group. Show that the set of all nonzero members of a field with the operation of multiplication (instead of addition) is a group.

31. A **commutative group** is a group for which the commutative law holds. Show that a field can be defined as an additive commutative group whose nonzero members form a multiplicative commutative group and for which the distributive law holds. (Cf. Ex. 30.)

32. Show that the set consisting of the distinct numbers 0 and 1 only, with the following addition and multiplication tables, is a field:

+	0	1
0	0	1
1	1	0

Addition

×	0	1
0	0	0
1	0	1

Multiplication

NOTE 1.   In an abstract group G (cf. Ex. 30) the member postulated by Axiom I(*iii*) is called the **identity member** of G, or the **identity element** of G, or, in brief, the **identity** of G.   In a general field $\mathscr{R}$ the member 0 of Axiom I(*iii*) is called the **additive identity** of $\mathscr{R}$ (as well as *zero*) and the member 1 of Axiom II(*iii*) is called the **multiplicative identity** of $\mathscr{R}$ (as well as *one* or *unity*).

NOTE 2.   In an abstract group G (cf. Ex. 30) the member postulated by Axiom I(*iv*) is called the **inverse** of x.   In a general field the member $-x$ of Axiom I(*iv*) is called the **additive inverse** of x (as well as the *negative* of x) and the member x^{-1} of Axiom II(*iv*) is called the **multiplicative inverse** of x (as well as the *reciprocal* of x).

2

Ordered Fields

♦♦

201. INTRODUCTION. POINT SET NOTATION

The axioms of a field, given in Chapter 1, yield only a relatively small portion of the properties of real numbers.   In terms of these axioms alone, for example, it is impossible to infer that there are infinitely many numbers (cf. Theorems V and VII, § 312).   The next set of axioms puts further restrictions on the number system and permits the introduction of an *order relation*. Since the language and terminology of sets will be important in the sequel, we introduce at this stage a few symbols from set theory.   For simplicity, members of sets will frequently be called *points*; when this is done no geometrical connotation should be attached to the word *point*.

Definition I.   *If a set or collection of objects is denoted by a letter, say A, the symbolic statement*

(1) $$a \in A$$

means that a is one of the objects of the set A.   This object a is called a **member** *or* **element** *or* **point** *of A.*

Example.   If $\mathscr{R}$ denotes the real number system or, for that matter, any field, then $0 \in \mathscr{R}$ and $1 \in \mathscr{R}$.

The equal sign $=$ between two sets, thus: $A = B$, means that the two sets A and B are identical collections of objects.   The statement $A \neq B$, then, means that at least one of the two sets contains a point that is not a member of the other set.

Definition II.   *The* **empty set**, *denoted* $\varnothing$, *is the set that has no members. That is, the statement* $a \in \varnothing$ *is never true.   A* **nonempty set** *is any set A containing at least one member; in symbols:* $A \neq \varnothing$.

Definition III.   *If A and B are sets, A is said to be a* **subset** *of B if and only if every member of A is a member of B.   This relationship is written:*

(2) $$A \subset B.$$

9

In other words, A is a subset of B if and only if:

(3) $x \in A$ implies $x \in B$.

The statement that A is a subset of B does not preclude the possibility that A may be empty, nor does it preclude the possibility that A may be identical with B. In fact, the following two relations are universally valid:

(4) $\varnothing \subset B$, $B \subset B$.

In a similar way, equality between two sets is equivalent to two inclusions:

(5) $A = B$ if and only if $A \subset B$ and $B \subset A$.

The statement that A is a **proper subset** of B means that A is a subset of B and that A is not equal to B: $A \subset B$ and $A \neq B$; in other words, that every point of A is a member of B and that there is at least one point of B that is not a member of A. In particular, the empty set is a proper subset of every nonempty set.

Two sets, A and B, are said to be **disjoint** if and only if they have no member in common; that is, the two statements $x \in A$ and $y \in B$ imply $x \neq y$.

202. AXIOMS OF ORDER

The axioms of the real number system $\mathscr{R}$ that are given in this chapter are expressed in terms of a certain special subset $\mathscr{P}$ of $\mathscr{R}$ (this subset $\mathscr{P}$ will soon be identified with the set of positive numbers). In general, *any* field having a subset $\mathscr{P}$ with the properties specified below is called an **ordered field**. The axioms of this chapter, then, state that *the real number system is an ordered field*. An arbitrary ordered field will be denoted $\mathscr{G}$ and, for simplicity, its members will be referred to as *numbers*.

IV. Order

There exists a set $\mathscr{P}$ of numbers such that:
  (*i*) *If $x \in \mathscr{P}$ and if $y \in \mathscr{P}$, then $x + y \in \mathscr{P}$.*
  (*ii*) *If $x \in \mathscr{P}$ and if $y \in \mathscr{P}$, then $xy \in \mathscr{P}$.*
  (*iii*) *For every number x exactly one of the following three statements is true:*
$$x \in \mathscr{P}; \quad x = 0: \quad -x \in \mathscr{P}.$$

In terms of the set $\mathscr{P}$ we now define the symbols of ordering:

Definition I. *The symbols $<$ and $>$* (read "is less than" and "is greater than," respectively) *are defined by the statements*

$$x < y \text{ if and only if } y - x \in \mathscr{P}.$$
$$x > y \text{ if and only if } x - y \in \mathscr{P}.$$

Each of the statements $x < y$ and $x > y$ is called an **inequality.**†

---

† Three books devoted to the general subject of inequalities are E. Beckenbach and R. Bellman, *An Introduction to Inequalities* (New York, Random House, 1961); G. H. Hardy, J. E. Littlewood, and G. Polya, *Inequalities*, Second edition (Cambridge University Press, 1952); and N. D. Kazarinoff, *Geometric Inequalities* (New York, Random House, 1961).

Definition II. *The symbols* $\leqq$ *and* $\geqq$ (read "is less than or equal to" and "is greater than or equal to," respectively) *are defined by the statements*

$$x \leqq y \text{ if and only if either } x < y \text{ or } x = y;$$
$$x \geqq y \text{ if and only if either } x > y \text{ or } x = y.$$

Each of the statements $x \leqq y$ *and* $x \geqq y$ *is called an* **inequality.**

It follows immediately from these definitions that the two statements (or inequalities) $x < y$ and $y > x$ are equivalent, and that the two statements (or inequalities) $x \leqq y$ and $y \geqq x$ are equivalent.

The *sense* of an inequality of the form $x < y$ or $x \leqq y$ is said to be the **reverse** of that of an inequality of the form $x > y$ or $x \geqq y$.

The simultaneous inequalities $x < y$, $y < z$ (that is, x is less than y and y is less than z) are usually written together, thus: $x < y < z$, and the simultaneous inequalities $x > y$, $y > z$ are usually written $x > y > z$. Similar interpretations are given the compound inequalities $x \leqq y \leqq z$ and $x \geqq y \geqq z$. In case $x < y < z$ or $x > y > z$, y is said to be **between** x and z.

Definition III. *A number* x *is* **positive** *if and only if* x *is a member of* $\mathscr{P}$. *A number* x *is* **negative** *if and only if* $-x$ *is positive.*

From Axiom IV(iii) we can infer that 0 is neither positive nor negative, and that 0 is the only number that is neither positive nor negative.

NOTE.   A subset A of an algebraic system S in which an operation of addition is defined is said to be **closed with respect to addition** if and only if whenever $x \in A$ and $y \in A$ it must follow that $x + y \in A$.   A similar definition applies to the concept of **closure with respect to multiplication, subtraction,** or **division.**   In terms of closure the first two axioms of this section can be reformulated:

IV(i): $\mathscr{P}$ is closed with respect to addition.

IV(ii): $\mathscr{P}$ is closed with respect to multiplication.

203. SOME SIMPLE CONSEQUENCES

In the following examples we illustrate a few properties of numbers that follow from the fact that the real number system is an ordered field.   The Exercises of § 204 contain several more properties for the reader to establish.

Example 1.   Prove that $x > 0$ if and only if x is positive.

Solution.   We start by using the definition of ordering (Definition I, § 202) in order to observe that the inequality $x > 0$ holds if and only if $x - 0 \in \mathscr{P}$.   By the definition of subtraction (§ 102), $x - 0 = x + (-0)$ and, by Exercise 2, § 104, $-0 = 0$, so that

$$x - 0 = x + (-0) = x + 0 = x.$$

Thus the inequality $x > 0$ holds if and only if $x \in \mathscr{P}$ or, in other words, if and only if x is positive.

Example 2. Establish the **transitive law** for both $<$ and $>$:

$$x < y, y < z \text{ imply } x < z,$$

$$x > y, y > z \text{ imply } x > z.$$

Solution. We establish the first implication by assuming that both inequalities on the left hold or, in other words, that both $y - x$ and $z - y$ are members of $\mathscr{P}$. As a consequence, by Axiom IV(*i*) the sum of these two members of $\mathscr{P}$ must also be a member of $\mathscr{P}$. By the formula of Exercise 23, § 104, this sum can be written:

$$(z - y) + (y - x) = z - x.$$

We are forced to conclude, since $z - x \in \mathscr{P}$, that $x < z$, as desired.

The second implication can be proved in a similar manner by assuming that both $x - y$ and $y - z$ are members of $\mathscr{P}$. An alternative proof is given by simply relabeling the first implication, with the letters x and z interchanged. The resulting implication:

$$z < y, y < x \text{ imply } z < x$$

is equivalent to the second.

Example 3. Establish the **law of trichotomy:** *For any x and y exactly one of the following three statements holds:*

(1) $$x < y, x = y, x > y.$$

Solution. Consider the number $z = y - x$. The first inequality of (1), $x < y$, holds if and only if $z = y - x$ is a member of $\mathscr{P}$. The equality $x = y$ holds if and only if $z = y - x$ is equal to 0, by elementary properties from Chapter 1. Finally, the second inequality of (1), $x > y$, holds if and only if $-z = -(y - x) = x - y$ is a member of $\mathscr{P}$. By Axiom IV(*iii*), as applied to the number $z = y - x$, exactly one of the three statements listed in (1) holds.

Example 4. Prove that addition of any number to both members of an inequality or subtraction of any number from both members of an inequality preserves the order relation:

$$x < y \text{ implies } x + z < y + z \text{ and } x - z < y - z,$$

$$x > y \text{ implies } x + z > y + z \text{ and } x - z > y - z,$$

$$x \leq y \text{ implies } x + z \leq y + z \text{ and } x - z \leq y - z,$$

$$x \geq y \text{ implies } x + z \geq y + z \text{ and } x - z \geq y - z.$$

Solution. The first two implications follow from elementary properties of Chapter 1, since

$$(y + z) - (x + z) = (y - z) - (x - z) = y - x,$$

$$(x + z) - (y + z) = (x - z) - (y - z) = x - y.$$

The last two implications are direct consequences of the first two by the definition of the symbols $\leq$ and $\geq$.

Example 5. Prove that the product of two negative numbers is positive and that the square of any nonzero number (that is, the product of the number and itself) is positive. Therefore the inequality $x^2 \geq 0$ (where $x^2 \equiv x \cdot x$) is true for all x. Conclude that unity is positive: $1 > 0$.

Solution. Let x and y be any two negative numbers. Then $-x$ and $-y$ are both positive, and so is their product $(-x)(-y)$. On the other hand, by Exercise 19, § 104, $xy = (-x)(-y)$, and the product xy of the two given negative numbers must also be positive. Since the square of any nonzero number is either the product of two positive numbers (the number and itself) or the product of two negative numbers (the number and itself), the result must be positive. Finally, since unity is its own square, it must be positive.

It follows from Example 5 that *the set $\mathscr{P}$ of positive numbers is nonempty*; that is, that there is at least one positive number. Furthermore, since $1 - 1 = 0$ and since 0 is not positive, it follows that *$\mathscr{P}$ is not closed with respect to subtraction.*

Example 6. Prove that the product of a positive number and a negative number is negative.

Solution. Let x be a positive number and y be a negative number. Then x and $-y$ are both positive numbers, and so is their product which, by Exercise 20, § 104, can be written:

$$x(-y) = -(xy).$$

Since, as a consequence, $-(xy)$ is a positive number, xy is negative, as we wished to show.

It follows from Examples 5 and 6 that *the product of two numbers is positive if and only if the two numbers are either both positive or both negative,* and that *the product of two numbers is negative if and only if one of the two numbers is positive and the other is negative.*

Example 7. Prove that the reciprocal of a positive number is positive and that the reciprocal of a negative number is negative.

Solution. Let x be a nonzero number. Then $xx^{-1} = 1 > 0$. But we have just seen that the product of two numbers is positive if and only if the two numbers are either both positive or both negative. Therefore x^{-1}, the reciprocal of x, is positive if x is positive, and it is negative if x is negative.

It follows from Example 7, in conjunction with Examples 5 and 6, that *the quotient of two numbers is positive if and only if the two numbers are either both positive or both negative,* and that *the quotient of two numbers is negative if and only if one of the two numbers is positive and the other is negative.* A portion of this conclusion can be rephrased: *The set $\mathscr{P}$ is closed with respect to division.*

204. EXERCISES

Prove the given statement, for an arbitrary ordered field.

1. $x < 0$ if and only if x is negative.

2. The sum of two negative numbers is negative.

3. If $2 \equiv 1 + 1$, then $2 > 1$ and $2 > 0$.

4. The equation $x^2 + 1 = 0$ has no real root; that is, no member of an ordered field can satisfy this equation.   (Cf. Ex. 28, § 104.)

5. The inequalities $0 < x < y$ imply the inequalities $0 < \dfrac{1}{y} < \dfrac{1}{x}$.

6. Multiplication or division of both members of an inequality by a positive number preserves the order relation: If $z > 0$, then $x < y$ implies $xz < yz$ and $x/z < y/z$, and $x \leq y$ implies $xz \leq yz$ and $x/z \leq y/z$; similarly for $>$ and $\geq$.

7. Multiplication or division of both members of an inequality by a negative number reverses the order relation: If $z < 0$, then $x < y$ implies $xz > yz$ and $x/z > y/z$, and $x \leq y$ implies $xz \geq yz$ and $x/z \geq y/z$; similarly for $>$ and $\geq$.

8. The simultaneous inequalities $a < b$ and $c < d$ imply $a + c < b + d$.   *Hint:* Add c to both members of the first inequality and b to both members of the second inequality, and use Examples 3 and 4, § 203.

9. The simultaneous inequalities $0 \leq a < b$ and $0 \leq c < d$ imply $ac < bd$. *Hint:* Consider the two cases: (i) $a = 0$ or $c = 0$, and (ii) a and c are both positive. (Cf. Ex. 8.)

10. If x and y are nonnegative numbers (that is, $x \geq 0$ and $y \geq 0$), then $x < y$ if and only if $x^2 < y^2$.   *Hint:* Cf. Ex. 28, § 104.

11. The transitive law holds for both $\leq$ and $\geq$: $x \leq y$, $y \leq z$ imply $x \leq z$; $x \geq y$, $y \geq z$ imply $x \geq z$.

12. The simultaneous inequalities $x \leq y$ and $y \leq x$ imply $x = y$.

13. If x is a fixed number satisfying the inequality $x < \epsilon$ for every positive number ϵ, then $x \leq 0$.   If x is a fixed number satisfying the inequality $x \leq \epsilon$ for every positive number ϵ, then $x \leq 0$.   Then, show that the inequality $x < \epsilon$ for every positive number ϵ does not imply the inequality $x < 0$.   *Hint:* If x were positive one could choose $\epsilon = \frac{1}{2}x$.   (Cf. Ex. 3.)

14. If x and y are distinct numbers, their **arithmetic mean** $\dfrac{x + y}{2}$ is between them. (Cf. Ex. 3.)

15. If a is an arbitrary number there exists a number x such that $x^2 > a$.   *Hint:* Consider the number x defined to be the larger of the two numbers 1 and $a + 1$ if $a \neq 0$, and equal to 1 if $a = 0$.

16. If a and b are nonzero numbers of opposite sign ($ab < 0$), the expression $ax^2 + b$ changes sign.   That is, there exists a number x such that $ax^2 + b > 0$ and there exists a number x such that $ax^2 + b < 0$.

17. If x and y are any two distinct numbers, then every number c between x and y is uniquely representable in the form

(1) $c = \alpha x + \beta y$, where $0 < \alpha < 1$, $0 < \beta < 1$, $\alpha + \beta = 1$.

Conversely, every number c of the form (1) is between x and y.   *Hint for existence:* Let $\alpha \equiv (y - c)/(y - x)$ and $\beta \equiv (c - x)/(y - x)$.

3

Natural Numbers and Mathematical Induction

301. INTRODUCTION

In any ordered field $\mathcal{G}$, the elements known as the *natural numbers* can be thought of as corresponding to the "counting numbers" of everyday usage: $1, 2, 3, 4, \cdots$. In slightly more precise terms, the number 1 is postulated to exist and to have certain simple properties, and the other "natural numbers" are defined one after the other, thus: $2 \equiv 1 + 1$, $3 \equiv 2 + 1$, $4 \equiv 3 + 1$, $5 \equiv 4 + 1$, and so forth. In Example 5, § 203, it is proved that 1 is a positive number: $1 > 0$. By Exercise 3, § 204, 2 is also a positive number and $2 > 1$. Similarly, 3 is a positive number and $3 > 2$, 4 is a positive number and $4 > 3$, etc. We thus obtain an increasing array of natural numbers such that

(1) $$0 < 1 < 2 < 3 < 4 < 5 < \cdots .$$

These introductory remarks seem simple enough—at least superficially—but they leave many questions unanswered. For example, what is the nature of the collection of *all* natural numbers? Are there infinitely many natural numbers, and what does this question really mean? Can there be a natural number less than 1, or between 2 and 3? What do the dots in (1) really mean? Suppose we subtract 1 from a natural number greater than 1; is the difference again a natural number? Does every nonempty set of natural numbers have a least member? Before attempting to answer such questions as these it is clearly necessary to formulate with precision exactly what the system of natural numbers, within an ordered field, really is. We continue to use the word *number* to mean an arbitrary member of a given ordered field $\mathcal{G}$.

It is important to keep in mind that when the real number system is described axiomatically as a complete ordered field, the familiar properties of the natural numbers—as set forth in this chapter—are all derivable from the axioms and need not be assumed independently. This is in contrast to

15

the constructive approach based on the Peano axioms for the natural numbers, as described in the Landau book cited in § 101. With the Peano approach, certain properties of the natural numbers are assumed as axioms, and the statements assumed as axioms in the present book are proved as theorems.

302. INDUCTIVE SETS

The main ideas in the crude description of the natural numbers given in the preceding section can be described very informally; (*i*) we start with 1, (*ii*) we keep adding 1 without stopping, and (*iii*) only the numbers obtained in this way are included in the set of natural numbers. In the first step toward our objective of a rigorous definition of the natural numbers we concentrate on (*i*) and (*ii*):

Definition. *An* **inductive set** *of numbers is a set A having the two properties:*
  (*i*) *The number 1 is a member of A:* $1 \in A$.
  (*ii*) *Whenever a number x belongs to A, the number x + 1 also belongs to A:* $x \in A$ *implies* $x + 1 \in A$.

Examples of inductive sets are easy to give, as shown below where $\mathscr{G}$ is an arbitrary ordered field and $\mathscr{P}$ is the subset of positive members of $\mathscr{G}$:

Example 1. Show that the set $\mathscr{G}$ is an inductive set.

Solution. Certainly 1 is a member of $\mathscr{G}$, and (*ii*) follows from the fact that $\mathscr{G}$ is closed with respect to addition, by Axiom I(*i*).

Example 2. Show that the set $\mathscr{P}$ is an inductive set.

Solution. By Example 5, § 203, $1 \in \mathscr{P}$, and (*ii*) follows from the fact that $\mathscr{P}$ is closed with respect to addition, by Axiom IV(*i*).

Example 3. Show that the set of all x such that $x \geq 1$ is an inductive set.

Solution. Clearly, 1 belongs to this set by definition. Assuming that x is a member of this set, we can conclude from the inequalities $x \geq 1 > 0$, by transitivity, that $x > 0$. Hence, if 1 is added to both members of this last inequality, by Example 4, § 203, $x + 1 > 1$ and $x + 1$ must also be a member of the set under consideration.

Example 4. Show that the set consisting of the number 1 together with all x such that $x \geq 2$ is an inductive set.

Solution. Again, 1 is a member by definition. We now assume that x is a member and seek to prove that $x + 1$ is too. There are two cases. In the first place, if $x = 1$, then $x + 1 = 2$, which is a member by definition. In the second place, if $x \geq 2$ then, by transitivity, $x > 1$ and (again by Example 4, § 203) if 1 is added to each member we obtain $x + 1 > 2$ and the desired conclusion that $x + 1$ is a member of the set under consideration.

303. THE SET OF NATURAL NUMBERS

Having introduced the concept of inductive set by pursuing the first two ideas of the first paragraph of § 302, we are now in a position to follow up on the third idea of that paragraph. The general aim is to exclude from the set of natural numbers everything extraneous to the bare minimum set of numbers obtained by adding 1 to itself repeatedly. To be more specific, we define the set of natural numbers to be the *smallest* of all inductive sets according to the following definition and theorem, all numbers under consideration being members of a single ordered field $\mathscr{G}$.

Definition. *A number n is a **natural number** if and only if n is a member of every inductive set.*

Theorem I. *The set $\mathscr{N}$ of all natural numbers is an inductive set. If $\mathscr{S}$ is an arbitrary inductive set of numbers, then $\mathscr{N}$ is a subset of $\mathscr{S}$:*

$$(1) \qquad\qquad \mathscr{N} \subset \mathscr{S}.$$

Proof. In order to show that $\mathscr{N}$ is an inductive set, we must show first that 1 is a member. Equivalently, by the preceding definition, we must show that 1 is a member of *every* inductive set. But this is true by definition of an inductive set. Next, we shall assume that n is a member of $\mathscr{N}$ and seek to prove that $n + 1$ must consequently also belong to $\mathscr{N}$. Accordingly, we assume that n belongs to every inductive set and wish to conclude that $n + 1$ also belongs to every inductive set. This conclusion is valid since, if $\mathscr{S}$ is an arbitrary inductive set, we infer that $n + 1$ belongs to $\mathscr{S}$ directly from the hypothesis that n belongs to $\mathscr{S}$. Hence $\mathscr{N}$ is an inductive set. The subset relation (1) follows immediately from the definition of $\mathscr{N}$: If n denotes an arbitrary member of $\mathscr{N}$, then n must be a member of every inductive set and therefore, in particular, of $\mathscr{S}$.

A corollary of this theorem is the fifth of the Peano axioms for the natural numbers (cf. § 101), the so-called *axiom of induction*:†

Theorem II. Axiom of Induction. *If $\mathscr{S}$ is a set of natural numbers with the two properties (i) $\mathscr{S}$ contains the number 1, and (ii) whenever $\mathscr{S}$ contains the natural number n it also contains the natural number $n + 1$, then $\mathscr{S}$ is the set of all natural numbers: $\mathscr{S} = \mathscr{N}$. In other words, the only inductive set of natural numbers is $\mathscr{N}$ itself.*

Proof. By assumption, $\mathscr{S} \subset \mathscr{N}$, and by the preceding theorem, $\mathscr{N} \subset \mathscr{S}$. That is, every member of each of the two sets is a member of the other, and

---

† This is also known as the *Principle of Induction*, or the *Principle of Finite Induction* if it is important to distinguish it from a similar principle in the theory of infinite sets known as the *Principle of Transfinite Induction*. Cf. E. J. McShane and T. A. Botts, *Real Analysis* (Princeton, D. Van Nostrand Co., Inc., 1959), Appendix II.

hence they consist of the same objects. This is the meaning of equality between the two sets: $\mathscr{S} = \mathscr{N}$.

From this result we infer immediately the following theorem (also known as the *Fundamental Theorem of Finite Induction*):

Theorem III. Fundamental Theorem of Mathematical Induction. *For every natural number n let P(n) be a proposition that is either true or false. If (i) P(1) is true and (ii) whenever the proposition P(n) is true for the natural number n the proposition P(n + 1) is also true, then P(n) is true for every natural number n.*

Proof. Let $\mathscr{S}$ be the set of natural numbers n for which $P(n)$ is true, and use the Axiom of Induction (Theorem II).

Two variants of the Fundamental Theorem of Mathematical Induction are stated in Exercise 33, § 308, and Exercise 12, § 506.

304. SOME PROPERTIES OF NATURAL NUMBERS

It will be our purpose in this section to establish the following seven basic properties of the natural numbers in an arbitrary ordered field $\mathscr{G}$:

    I. *The natural numbers are positive.*

    II. *If n is a natural number, then $n \geq 1$; that is, 1 is the least natural number.*

    III. *If m and n are natural numbers, then m + n is a natural number; that is, $\mathscr{N}$ is closed with respect to addition.*

    IV. *If m and n are natural numbers, then mn is a natural number; that is, $\mathscr{N}$ is closed with respect to multiplication.*

    V. *If m and n are natural numbers and if $m < n$, then $n - m$ is a natural number.*

    VI. *If m is a natural number, there is no natural number n such that $m < n < m + 1$; that is, there is no natural number between the* **consecutive** *natural numbers m and m + 1.*

    VII. **Well-ordering Principle.**† *Every nonempty set of natural numbers (that is, every set of natural numbers that contains at least one member) contains a least or smallest member.*

Proof of I. By Example 2, § 302, the set $\mathscr{P}$ is an inductive set, and hence (Theorem I, § 303) $\mathscr{N} \subset \mathscr{P}$. But the inclusion statement $\mathscr{N} \subset \mathscr{P}$ is equivalent to statement I.

Proof of II. Let $\mathscr{S}$ denote the set of all numbers x such that $x \geq 1$. By Example 3, § 302, $\mathscr{S}$ is an inductive set, and hence $\mathscr{N} \subset \mathscr{S}$.

Proof of III. For an arbitrary given natural number m, let $\mathscr{S}_m$ denote the set of all numbers x such that $m + x$ is a natural number. Then, since

---

† For a discussion of well-ordered sets in general, see the McShane and Botts book cited in the footnote of the preceding section.

$m + 1$ is a natural number, 1 is a member of $\mathscr{S}_m$. Furthermore, if we assume that x is a member of $\mathscr{S}_m$ we can conclude that $x + 1$ must also be a member of $\mathscr{S}_m$ as follows: assuming that $m + x$ is a natural number entails the further assumption that $(m + x) + 1$ is a natural number. But, by the associative law I(ii), this means that $m + (x + 1)$ is a natural number and hence that $x + 1$ is a member of $\mathscr{S}_m$. Therefore $\mathscr{S}_m$ is an inductive set, and thus $\mathscr{N} \subset \mathscr{S}_m$. With this inclusion established the proof of III is complete.

Proof of IV. For an arbitrary given natural number m, let $\mathscr{S}_m$ denote the set of all numbers x such that mx is a natural number. Clearly, 1 is a member of $\mathscr{S}_m$. Furthermore, whenever x is a member of $\mathscr{S}_m$, $x + 1$ must also be a member of $\mathscr{S}_m$ since, by the distributive law of § 102, $m(x + 1) = mx + m$, and this last number has the form of the sum of two natural numbers which, by III, must again be a natural number. As with the proof of III, therefore, $\mathscr{S}_m$ is an inductive set, $\mathscr{N} \subset \mathscr{S}_m$, and the proof is complete.

Proof of V for $m = 1$. (The proof for $m > 1$ follows the proof of VI.) Assume there exists a natural number n_0 greater than 1 such that $n_0 - 1$ is *not* a natural number. Let $\mathscr{S}$ denote the set obtained by deleting from $\mathscr{N}$ the number n_0; that is, $\mathscr{S}$ consists of *all* natural numbers *except* n_0. We shall now proceed to show that $\mathscr{S}$ is an inductive set. In the first place, 1 is a member of $\mathscr{S}$ since 1 is a natural number, and 1 was not deleted from $\mathscr{N}$ in the formation of $\mathscr{S}$ since $n_0 \neq 1$. Next, we assume that n is a member of $\mathscr{S}$ and wish to conclude that $n + 1$ is also a member of $\mathscr{S}$. This must be true for the following reason: since n is a member of $\mathscr{S}$, n is a natural number and hence so is $n + 1$. Therefore, if $n + 1$ were *not* a member of $\mathscr{S}$, this number $n + 1$ would necessarily be the number n_0 deleted from $\mathscr{N}$ in the formation of $\mathscr{S}$: $n + 1 = n_0$. It follows that $n = n_0 - 1$, and a contradiction is obtained, since n is a natural number, whereas $n_0 - 1$ is not. This contradiction establishes the fact that whenever n is a member of $\mathscr{S}$, so is $n + 1$. Therefore $\mathscr{S}$ is an inductive set, as we wished to show. From this fact, finally, we can conclude from Theorem I, § 303, that $\mathscr{N} \subset \mathscr{S}$. But n_0 is a member of $\mathscr{N}$ but not of $\mathscr{S}$. With this final contradiction the proof of V for the case $m = 1$ is complete.

Proof of VI for $m = 1$. This follows from Example 4, § 302, since the set $\mathscr{S}$ of that example is an inductive set containing no numbers between 1 and 2. Since $\mathscr{N} \subset \mathscr{S}$, there can be no natural number between 1 and 2.

Proof of VI. Let $\mathscr{S}$ be the set of all natural numbers m such that there is no natural number n between m and $m + 1$. By the proof immediately preceding, 1 is a member of the set $\mathscr{S}$ just defined. Now assume that m is a member of $\mathscr{S}$ (that is, that there is no natural number between m and $m + 1$). We wish to prove that $m + 1$ is a member of $\mathscr{S}$ (that is, that there is no natural number between $(m + 1)$ and $(m + 1) + 1 = m + (1 + 1) = m + 2)$. Assume that there *is* such a natural number, and denote it by n:

(1) $$m + 1 < n < m + 2.$$

Since $m > 0$ (by I), $m + 1 > 1$ (adding 1 to each member—cf. Example 4, § 203) and hence, by transitivity, $n > 1$. Therefore, by the preceding proof of V for $m = 1$, $n - 1$ is a natural number. Finally, subtracting 1 from all three members of (1) (cf. Example 4, § 203), we obtain the inequalities $m < n - 1 < m + 1$. This contradicts the assumption that m is a member of $\mathscr{S}$ (since the natural number $n - 1$ is between m and $m + 1$). Therefore $\mathscr{S}$ is an inductive set of natural numbers, and hence $\mathscr{S} = \mathscr{N}$ (cf. § 303), and the proof is complete.

Proof of V. Let $\mathscr{S}$ be the set of all natural numbers m such that for every natural number n greater than m the number $n - m$ is a natural number. By the proof of the special case $m = 1$ (preceding), 1 is a member of $\mathscr{S}$. Assuming that m is a member of $\mathscr{S}$, we shall show that $m + 1$ is also a member of $\mathscr{S}$. That is, we assume that whenever n is a natural number such that $n > m$, then $n - m$ is a natural number, and seek to prove that whenever k is a natural number such that $k > m + 1$, then $k - (m + 1)$ is a natural number. Since $m > 0$ (by I), $m + 1 > 1$ (adding 1 to each member—cf. Example 4, § 203) and hence, by transitivity, if $k > m + 1$, then $k > 1$. Therefore, by the proof of V for $m = 1$ (preceding), $k - 1$ is a natural number. From the inequality $k > m + 1$ we can infer by subtraction of 1 (Example 4, § 203) that $k - 1 > m$. By assumption, then, since $k - 1$ is a natural number greater than m, the difference, $(k - 1) - m$, is a natural number. But by elementary properties of numbers developed in Chapter 1, $(k - 1) - m = [k + (-1)] + (-m) = k + [(-1) + (-m)] = k + [-(m + 1)] = k - (m + 1)$. With the desired conclusion that $k - (m + 1)$ is a natural number, the proof is complete.

Proof of VII. Assume that there is a nonempty set A of natural numbers that has no least member. Let $\mathscr{S}$ be the set of all natural numbers n having the property that $n < a$ for every member a of A. In the first place, 1 is a member of $\mathscr{S}$ since the inequality $1 \leqq a$ holds for every member a of A (by II), and if 1 were a member of A, 1 would be the *least* member of A, which is assumed not to exist. That is, in the inequality $1 \leqq a$, equality cannot hold, and hence $1 < a$ must hold for every a in A. Now assuming that n is a member of $\mathscr{S}$, we wish to show that $n + 1$ is also a member of $\mathscr{S}$. That is, we assume that the inequality $n < a$ holds for every a of A. But since *no* member a of A can satisfy the inequality $n < a < n + 1$ (by (VI) it follows that *every* member a of A must satisfy the inequality $n + 1 \leqq a$. As before, equality is ruled out since, if any member a of A were equal to $n + 1$, then this member would be the *least* member of A, which does not exist. In other words, the inequality $n + 1 < a$ must hold for all members a of A, and $n + 1$

is a member of $\mathscr{S}$, as was to be shown.   We can now conclude (cf. § 303) that $\mathscr{S} = \mathscr{N}$.   This means that for every natural number n and every member a of A the inequality $n < a$ must hold.   In particular, since $A \subset \mathscr{N}$, the inequality $a < a$ must hold for every a of A.   With this final contradiction, since A is not empty, the proof of VII is complete.

NOTE.   The set $\mathscr{N}$ of natural numbers is *not* closed with respect to division.   This means that, although for *some* natural numbers m and n the number m/n *may* be a natural number (as when 6 is divided by 2), *there exist* natural numbers m and n such that m/n is *not* a natural number.   A single example suffices to demonstrate this fact, and as simple an example as any is $m = 1$ and $n = 2$, since the number $1/2$, being less than 1, cannot be a natural number (cf. Property II of this section).

305. GENERAL ASSOCIATIVE, COMMUTATIVE, AND DISTRIBUTIVE LAWS

The associative laws for addition and multiplication, § 102:

$$(1) \qquad x + (y + z) = (x + y) + z, \qquad x(yz) = (xy)z,$$

state that any two sums or products of three numbers x, y, and z (where two or more of these numbers may be identical) in the same order are equal regardless of the manner in which the terms or factors are grouped by parentheses.

A similar formulation for four numbers, w, x, y, and z, for the multiplicative case would take the form of equating the following five products:

$$(2) \qquad (wx)(yz) = ((wx)y)z = w(x(yz)) = (w(xy))z = w((xy)z),$$

with increasing complexity accompanying an increase in the number of factors.   Before extending our study of the associative laws to n terms or n factors, for an arbitrary natural number n, we must turn our attention to some essential preliminaries.

In the first place, we must agree on what is meant by the two expressions "the n points $x_1, x_2, \cdots, x_n$" and "a set of n things."

Definition I.   *If to each natural number k satisfying the inequalities $1 \leq k \leq n$, there corresponds exactly one object or point x_k, we speak of the set or collection of points, denoted $\{x_k\}$, $k = 1, 2, \cdots, n$, or $\{x_1, x_2, \cdots, x_n\}$, as* **the set consisting of the n points** $x_1, x_2, \cdots, x_n$, *whether any two or more of these points are the same or not.   If no two are the same, we shall call the points* **distinct** *and speak of the set of points $x_1, x_2, \cdots, x_n$ as* **a set of n distinct points**.

We shall now restrict the points or objects under consideration to numbers (or, more generally, to members of *an arbitrary field*).   We wish to formulate what is meant by a sum or product of the n numbers $x_1, x_2, \cdots, x_n$, for any natural number n.   For simplicity of notation we restrict consideration to products only (the details for sums being entirely similar).

Definition II. *Products of the n numbers* $x_1, x_2, \cdots, x_n$ *in the given order are defined as follows: If* $n = 1$, *the product is* x_1, *and is unique. If* $n = 2$, *the product is* $x_1 x_2$, *and is unique. If* $n = 3$, *there are two products,* $x_1(x_2 x_3)$ *and* $(x_1 x_2)x_3$ *which, by the associative law of* § *102, are equal. Assuming that all possible products of any n numbers* $x_1, x_2, \cdots, x_n$ *in the given order have been defined, the products of any* $n + 1$ *numbers* $x_1, x_2, \cdots, x_n, x_{n+1}$ *are defined to be all possible products in given order of n numbers of the form* $y_1, y_2, \cdots y_n$ *where* $y_k = x_k x_{k+1}$ *for some k such that* $1 \leq k \leq n$, $y_m = x_m$ *for all m such that* $1 \leq m < k$, *and* $y_m = x_{m+1}$ *for all m such that* $k < m \leq n$.

Example 1. The possible products of four numbers x_1, x_2, x_3, and x_4 are the products of the three numbers $x_1 x_2$, x_3, and x_4, the products of the three numbers x_1, $x_2 x_3$, and x_4, and the products of the three numbers x_1, x_2, and $x_3 x_4$. (These total to five distinct product forms, as shown in (2), instead of six since one of the products of $x_1 x_2$, x_3, and x_4 is identical with one of the products of x_1, x_2, and $x_3 x_4$.)

Theorem I. *If n is any natural number and if* $x_1, x_2, \cdots, x_n$ *are any n numbers, the sums and the products of these n numbers in the given order exist.*

Proof. Let $P(n)$ be the statement that for any n numbers $x_1, x_2, \cdots, x_n$ the sums and products exist. Then $P(1)$ is true by definition. Furthermore, by the manner in which sums and products are defined, whenever $P(n)$ is true, $P(n + 1)$ is also true. Therefore, by the Fundamental Theorem of Mathematical Induction (Theorem III, § 303), $P(n)$ is true for every natural number n.

Theorem II. *For any natural number* $n \geq 2$, *any product of the n numbers* $x_1, x_2, \cdots, x_n$ *in that order must have the form ab, where a is a product (for some k such that* $1 \leq k \leq n - 1$) *of the k numbers* $x_1, x_2, \cdots, x_k$ *and b is a product of the* $n - k$ *numbers* $x_{k+1}, \cdots, x_n$.

Proof. Let $P(1)$ be any true proposition and, for $n \geq 2$, let $P(n)$ be the proposition asserted in the theorem. Then $P(2)$ is trivially true. For any $n \geq 2$ assume that $P(n)$ is true, and consider any product of $n + 1$ numbers $x_1, x_2, \cdots, x_{n+1}$. By Definition II, this product must be a product of n numbers of the form $y_1, y_2, \cdots, y_n$, where $y_k = x_k x_{k+1}$ for some k such that $1 \leq k \leq n$, $y_m = x_m$ for all m such that $1 \leq m < k$, and $y_m = x_{m+1}$ for all m such that $k < m \leq n$. By the induction assumption, the product of the y's must have the form ab, where a is a product of the first j y's and b is a product of the last $n - j$ y's. Again by Definition II, the two numbers a and b have the desired form in terms of the x's. By the Fundamental Theorem of Mathematical Induction (§ 303), $P(n)$ is true for every natural number n.

We are now ready to state a general form for the principle of associativity:

Theorem III. General Associative Laws. *Any two sums (products) of the n numbers* $x_1, x_2, \cdots, x_n$ *in the given order are equal regardless of the manner in which the terms (factors) are grouped by parentheses.*

Example 2. Let $a = x_1((x_2x_3)(x_4x_5))$ and $b = (((x_1x_2)x_3)x_4)x_5$. We shall show that $a = b$ by using the associative law of II, § 102, to transform a step by step into b. A similar sequence of steps would transform any product of the five numbers $x_1, x_2, \cdots, x_5$ into the "standard" product b, and hence justify the theorem for $n = 5$. We start by thinking of the products x_2x_3 and x_4x_5 as single numbers and we use the associative law to write $a = (x_1(x_2x_3))(x_4x_5)$. Repeating this method we have: $a = ((x_1x_2)x_3)(x_4x_5) = (((x_1x_2)x_3)x_4)x_5 = b$.

We now give a detailed proof of Theorem III, restricting considerations to the multiplicative form for simplicity of notation.

Proof. Let $P(n)$ be the proposition: "Any product of the m numbers $x_1, x_2, \cdots, x_m$, in that order, is equal to the special product $((\cdots((x_1x_2)x_3) \cdots)x_{m-1})x_m$, whenever $m \leq n$. For $n = 1$ and $n = 2$ the proposition is trivial, and for $n = 3$ it follows from the associative law of II, § 102, $x_1(x_2x_3) = (x_1x_2)x_3$. Assume now the truth of $P(n)$, for a fixed n, and consider any possible form for the product of the $n + 1$ numbers $x_1, x_2, \cdots, x_{n+1}$, in that order. By Theorem II, such a product must have the form ab, where a and b are products of at most n of the x's. By the induction assumption, each of these two factors can be rewritten, if necessary, in the form $a = yx_k$ and $b = zx_{n+1}$, where y is either 1 (in case $k = 1$) or a product of the factors $x_1, x_2, \cdots, x_{k-1}$, and z is either 1 (in case $k = n$) or a product of the factors $x_{k+1}, \cdots, x_n$. By the associative law of II, § 102, $ab = (yx_k)(zx_{n+1}) = ((yx_k)z)x_{n+1}$. Again using the induction hypothesis, we can write the product $(yx_k)z$ in the special form $((\cdots((x_1x_2)x_3) \cdots)x_{n-1})x_n$. This fact, with the aid of the Fundamental Theorem of Mathematical Induction (Theorem III, § 303), establishes the truth of $P(n)$ for every natural number n. Finally, since any two products of n numbers in a given order are equal to the same special product, they must be equal to each other, and the proof is complete.

NOTE. By the associative laws any sum or product of n numbers, in a given order, can be written without parentheses, thus:

$$(3) \qquad\qquad x_1 + x_2 + \cdots + x_n \quad \text{and} \quad x_1x_2 \cdots x_n.$$

Theorem IV. General Commutative Laws. *Any two sums (products) of the n numbers $x_1, x_2, \cdots, x_n$ are equal regardless of the order of the terms (factors).*

Example 3. Let $a = x_4x_1x_5x_2x_3$ and $b = x_1x_2x_3x_4x_5$. We shall show that $a = b$ by using the commutative law of II, § 102, to transform a step by step into b. We first bring x_1 to the left-hand end: $a = (x_4x_1)(x_5x_2x_3) = (x_1x_4)(x_5x_2x_3) = x_1x_4x_5x_2x_3$. Next we take care of x_2: $a = x_1x_4(x_5x_2)x_3 = x_1x_4(x_2x_5)x_3 = x_1(x_4x_2)x_5x_3 = x_1(x_2x_4)x_5x_3 = x_1x_2x_4x_5x_3$. Finally, after x_3 is moved two steps to the left, the form b is reached.

As with the proof of Theorem III, we restrict ourselves in the proof of Theorem IV to the multiplicative form for simplicity of notation.

Proof. Let $P(n)$ be the proposition: "Any two products of n numbers are equal regardless of the order of the factors." For $n = 1$ the proposition is trivial and for $n = 2$ it follows from the commutative law of II, § 102: $x_2 x_1 = x_1 x_2$. Assume now the truth of $P(n)$ for a particular n and consider any possible product of the $n + 1$ numbers $x_1, x_2, \cdots, x_{n+1}$. This product must have the form $x x_{n+1} y$, where x is either 1 or the product of some of the x's, and y is either 1 or the product of some of the x's. By the commutative and associative laws of § 102, $x x_{n+1} y = x(x_{n+1} y) = x(y x_{n+1}) = (xy) x_{n+1}$. The product xy contains the n factors $x_1, x_2, \cdots, x_n$ which, by the induction assumption that $P(n)$ is true, can be rearranged according to the order of the subscripts. Therefore $P(n + 1)$ follows from $P(n)$, and application of the Fundamental Theorem completes the proof.

The general distributive law is now easy to formulate and prove:

Theorem V. General Distributive Law. *If x is any number and if y_1, $y_2, \cdots, y_n$ are any n numbers,*

(4) $$x(y_1 + y_2 + \cdots + y_n) = xy_1 + xy_2 + \cdots + xy_n.$$

Proof. Let $P(n)$ be the proposition (4). $P(1)$ is a triviality, and $P(2)$ is true by the distributive law III, § 102. We wish to show now that the truth of (4) for a particular natural number n implies the truth of $P(n + 1)$:

(5) $x(y_1 + y_2 + \cdots + y_n + y_{n+1}) = xy_1 + xy_2 + \cdots + xy_n + xy_{n+1}.$

By using the distributive law of § 102 and the assumption (4), we can rewrite the left-hand member of (5) as follows:

$$x[(y_1 + y_2 + \cdots + y_n) + y_{n+1}] = x(y_1 + \cdots + y_n) + xy_{n+1}$$
$$= (xy_1 + \cdots + xy_n) + xy_{n+1}.$$

Since this last expression is equal to the right-hand member of (5), the truth of $P(n)$, or (4), is established for all natural numbers n by the Fundamental Theorem of Mathematical Induction.

306. MORE ABOUT MATHEMATICAL INDUCTION

The establishment of a proposition $P(n)$ for every natural number n by showing that $P(1)$ is true and that $P(n)$ implies $P(n + 1)$—and invoking the aid of the Fundamental Theorem of Mathematical Induction (Theorem III, § 303)—is called a **proof by mathematical induction**. This method was used in § 305 to prove the general associative, commutative, and distributive laws. In this and in following sections we give further examples of proofs by mathematical induction.

Example 1. Prove that, in any ordered field, if $x_1 < x_2, x_2 < x_3, \cdots, x_{n-1} < x_n$ (usually written $x_1 < x_2 < \cdots < x_n$), then $x_1 < x_n$.

Solution. Let $P(1)$ be any true proposition, and for $n \geq 2$, let $P(n)$ be the proposition whose proof is requested. Then $P(2)$ is trivially true. For any $n \geq 2$ assume that $P(n)$ is true and that $x_1, x_2, \cdots, x_{n+1}$ are $n + 1$ points such that $x_k < x_{k+1}$ for $k = 1, 2, \cdots, n$. By the induction assumption, $x_1 < x_n$, and therefore by the transitive law for $<$ (Example 2, § 203), since $x_n < x_{n+1}$, we conclude that $x_1 < x_{n+1}$. In other words, for $n \geq 2$, $P(n)$ implies $P(n + 1)$. By the Fundamental Theorem of Mathematical Induction (§ 303), $P(n)$ is true for every natural number n.

Example 2. Prove that, in any ordered field, the sum and product of n natural numbers are natural numbers.

Solution. Let $P(n)$ be the proposition that if $m_1, m_2, \cdots, m_n$ are natural numbers then so are their sum and product. Then $P(1)$ is trivially true. Assuming now the truth of $P(n)$ for a particular natural number n, we wish to establish $P(n + 1)$, or that the sum and product of the natural numbers $m_1, m_2, \cdots, m_{n+1}$ are natural numbers. To be specific we consider the sum of these natural numbers, which (by the general associative and commutative laws) can be written in the form $a + m_{n+1}$, where $a = m_1 + m_2 + \cdots + m_n$. By the induction assumption, a is a natural number. Therefore, by III, § 304, $a + m_{n+1}$ is also a natural number, and $P(n + 1)$ is true. By the Fundamental Theorem of Mathematical Induction (§ 303), $P(n)$ is true for every natural number n.

Example 3. Prove that, in an arbitrary field, any product of n nonzero numbers is nonzero.

Solution. The details are nearly the same as those given for Example 2, for the case of products, the property of being a natural number being replaced by that of being a nonzero number. The critical property needed is that given in Example 4, § 103: The product of two nonzero numbers is nonzero.

Example 4. Prove that, in any ordered field $\mathscr{G}$, any set consisting of the n points $x_1, x_2, \cdots, x_n$ contains a least member and a greatest member; that is, the set $x_1, x_2, \cdots, x_n$ contains members x_i and x_j such that for every member $x_k (k = 1, 2, \cdots, n)$: $x_i \leq x_k \leq x_j$. These two numbers x_i and x_j are written:

$$x_i = \min (x_1, x_2, \cdots, x_n) = \min_{k=1}^{n} (x_k)$$

$$x_j = \max (x_1, x_2, \cdots, x_n) = \max_{k=1}^{n} (x_k).$$

Solution. Only the details for x_i will be given (those for x_j are entirely similar except for the sense of the inequalities, and can be supplied by the reader). Accordingly, let $P(n)$ be the statement that any set of the form $x_1, x_2, \cdots, x_n$ contains a member x_i such that $x_i \leq x_k$ for every member x_k. The case $P(1)$ is trivially true. Assuming the truth of $P(n)$, for a fixed natural number n, we wish to establish the truth of $P(n + 1)$. Let $x_1, x_2, \cdots, x_{n+1}$ be an arbitrary set consisting of $n + 1$ members of $\mathscr{G}$. By the induction assumption, there exists a member x_m of the set $x_1, x_2, \cdots, x_n$ such that $x_m \leq x_k$ for every k for which $1 \leq k \leq n$. There are two cases to consider: (*i*) If $x_m \leq x_{n+1}$, then $x_m \leq x_k$ for every k for which $1 \leq k \leq n + 1$, and we can choose $i = m$ to obtain the desired $x_i = x_m$. (*ii*) If

$x_{n+1} < x_m$, then we can choose $i = n + 1$ to obtain $x_i < x_m \leq x_k$ for every k for which $1 \leq k \leq n$, and therefore $x_i \leq x_k$ for every k for which $1 \leq k \leq n + 1$, as desired. An application of the Fundamental Theorem of Mathematical Induction (§ 303) completes the proof.

307. LAWS OF EXPONENTS

If n is any natural number, the **nth power** of a number x, denoted x^n, is defined to be the product of the n numbers $x_1, x_2, \cdots, x_n$, where every x_k is equal to x, $k = 1, 2, \cdots, n$. Thus, $x^2 = x \cdot x$, $x^3 = x \cdot x \cdot x$, etc. The number n is called the **exponent** of the power x^n.

The familiar laws of exponents, as set forth in the following theorem, are good examples of propositions to be proved by mathematical induction. Extensions of these laws, where the exponents are not restricted to the natural numbers, are considered in later sections (§§ 503, 1006, 1104).

Theorem. *If x and y are any numbers (or, more generally, members of an arbitrary field) and if m and n are any natural numbers, subject to possible restriction as specified, then:*

(i) $$x^m x^n = x^{m+n};$$

(ii) $$\text{if } x \neq 0, \quad \frac{x^m}{x^n} = \begin{cases} x^{m-n} & \text{if } m > n, \\ \dfrac{1}{x^{n-m}} & \text{if } m < n; \end{cases}$$

(iii) $$(x^m)^n = x^{mn};$$

(iv) $$(xy)^n = x^n y^n;$$

(v) $$\text{if } y \neq 0, \quad \left(\frac{x}{y}\right)^n = \frac{x^n}{y^n}.$$

Proof. (i): Let m be a fixed, but arbitrary, natural number, and let $P(n)$ be the proposition (i). Then $P(1)$ is true since, if we let $x_k \equiv x$ for $1 \leq k \leq m + 1$, then the left-hand member of (i) is $(x_1 x_2 \cdots x_m)x_{m+1}$ whereas the right-hand member is $x_1 x_2 \cdots x_m x_{m+1}$, and these two products are equal by the general associative law. Assuming the truth of $P(n)$ for a specific natural number n, we wish to infer that of $P(n + 1)$:

(1) $$x^m x^{n+1} = x^{m+n+1}.$$

By the special case $P(1)$, just established, the left-hand member of (1) can be written

$$x^m(x^n x) = (x^m x^n)x$$

which, by the induction assumption and the special case $P(1)$, is equal to

$$(x^{m+n})x = x^{(m+n)+1} = x^{m+n+1},$$

which is the right-hand side of (1).   By the Fundamental Theorem (§ 303), the proof of (*i*) is complete.

(*ii*): If $m > n$, then $m - n$ and n are natural numbers whose sum is m, so that by (*i*):

$$x^{m-n}x^n = x^m.$$

Division by x^n (which, by Example 3, § 306, is nonzero) gives the desired formula of (*ii*).   If $m < n$, then $n - m$ and m are natural numbers whose sum is n, so that by (*i*):

$$x^{n-m}x^m = x^n.$$

Division of both members by $x^{n-m}x^n$ gives the second formula of (*ii*).

(*iii*): Let m be a fixed, but arbitrary, natural number, and let $P(n)$ be the proposition (*iii*).   Then $P(1)$ is trivially true.   Assuming the truth of $P(n)$, for a specific natural number n, we wish to infer that of $P(n + 1)$:

(2) $$(x^m)^{(n+1)} = x^{m(n+1)}.$$

By part (*i*), already established, and with the use of the assumed proposition (*iii*) for the natural number n, we have for the left-hand member of (2):

$$(x^m)^n(x^m)^1 = (x^{mn})x^m = x^{mn+m}.$$

This last power of x has exponent equal to that of the right-hand member of (2), as desired.

(*iv*): If $P(n)$ is the proposition (*iv*), $P(1)$ is trivially satisfied.   Assuming the truth of $P(n)$ for a specific natural number n, we wish to infer that of $P(n + 1)$:

(3) $$(xy)^{n+1} = x^{n+1}y^{n+1}.$$

By the induction assumption that $P(n)$ holds, and with the aid of part (*i*) and the associative and commutative laws, we can rewrite the left-hand member of (3):

$$(xy)^{n+1} = (xy)^n(xy) = (x^ny^n)(xy) = (x^nx)(y^ny) = x^{n+1}y^{n+1},$$

and (3) is established.

(*v*): If $y \neq 0$, we can use (*iv*) to write:

$$\left(\frac{x}{y}\right)^n y^n = \left(\frac{x}{y}y\right)^n = x^n.$$

Division of both extreme terms by the nonzero quantity y^n completes the proof of (*v*), and therefore that of the theorem.

Example 1.   Use mathematical induction to establish the formula

(1) $$1^2 + 3^2 + 5^2 + \cdots + (2n - 1)^2 = \tfrac{1}{3}n(4n^2 - 1)$$

for every natural number n.

Solution.   Let $P(n)$ be the proposition (1).   Direct substitution shows that $P(1)$ is true.   We wish to show that whenever $P(n)$ is true for a particular natural number

n it is also true for the natural number $n + 1$. Accordingly, we assume (1) and wish to establish

(2) $$1^2 + 3^2 + 5^2 + \cdots + (2n - 1)^2 + (2n + 1)^2$$
$$= \tfrac{1}{3}(n + 1)[4(n + 1)^2 - 1].$$

On the assumption that (1) is correct (for a particular value of n), we can rewrite the left-hand member of (2) by grouping:

$$[1^2 + 3^2 + \cdots + (2n - 1)^2] + (2n + 1)^2 = \tfrac{1}{3}n(4n^2 - 1) + (2n + 1)^2.$$

Thus verification of (2) reduces to verification of

(3) $$\tfrac{1}{3}n(4n^2 - 1) + (2n + 1)^2 = \tfrac{1}{3}(n + 1)(4n^2 + 8n + 3),$$

which, in turn, is true (divide by 3) by virtue of

(4) $$4n^3 - n + 3(4n^2 + 4n + 1) = (4n^3 + 8n^2 + 3n) + (4n^2 + 8n + 3).$$

By the Fundamental Theorem of Mathematical Induction, (1) is true for all natural numbers n.

Example 2. Establish the inequality $m^n > n$ for all natural numbers m and n, where $m > 1$.

Solution. For an arbitrary but fixed natural number m greater than 1, let $P(n)$ be the statement of inequality $m^n > n$ for the natural number n. The particular case $P(1)$ is true by the assumption $m > 1$. Assuming the truth of $P(n)$, or $m^n > n$, for the natural number n, we wish to establish the truth of $P(n + 1)$, or $m^{n+1} > n + 1$. By the induction assumption we have $m^{n+1} = m^n \cdot m > n \cdot m \geq n(1 + 1) = n + n \geq n + 1$ (cf. Ex. 6, § 204), and $P(n + 1)$ follows. With an application of the Fundamental Theorem of Mathematical Induction the proof is complete.

308. EXERCISES

In the following exercises the word *number* means a member of an ordered field, unless specification otherwise is made.

1. Prove that the sum and product of n positive numbers are positive.

2. In an arbitrary field, if $-x_1 - x_2 - \cdots - x_n$ is defined to be $(-x_1) + (-x_2) + \cdots + (-x_n)$, prove that $-(x_1 + x_2 + \cdots + x_n) = -x_1 - x_2 - \cdots - x_n$.

3. In an arbitrary field, if $x_1, x_2, \cdots, x_n$ are nonzero numbers, prove that $(x_1 x_2 \cdots x_n)^{-1} = x_1^{-1} x_2^{-1} \cdots x_n^{-1}$.

4. If x^0 is defined to be equal to 1 for any number x, and if $\mathcal{N}_0$ denotes the **extended natural number system** consisting of 0 and all natural numbers, prove that the laws of exponents as stated in the Theorem, § 307, remain valid if m and n are permitted to be arbitrary members of $\mathcal{N}_0$. In part (*ii*) consider also the case $m = n$. (Cf. the footnote, § 503).

5. If n is any natural number, prove: $1^n = 1$, $(-1)^{2n} = 1$, $(-1)^{2n+1} = -1$.

6. If $a > 1$ and if m and n are natural numbers, prove that the inequality $m < n$ is equivalent to the inequality $a^m < a^n$. If $0 < a < 1$ and if m and n are natural numbers, prove that the inequality $m < n$ is equivalent to the inequality $a^m > a^n$.

7. If $a > 0$, if $a \neq 1$, and if m and n are natural numbers, prove that $a^m = a^n$ if and only if $m = n$. (Cf. Ex. 6.)

8. If a and b are nonnegative members of an ordered field and if n is a natural number, prove that the inequality $a < b$ is equivalent to the inequality $a^n < b^n$, and that $a^n = b^n$ if and only if $a = b$.

9. If a and b are arbitrary members of an ordered field and if n is a natural number, prove that the inequality $a < b$ is equivalent to the inequality $a^{2n+1} < b^{2n+1}$.

10. In an arbitrary field, prove that

$$x^n - y^n = (x - y)(x^{n-1} + x^{n-2}y + \cdots + xy^{n-2} + y^{n-1}),$$

where n is a natural number. *Hint:* $x^{n+1} - y^{n+1} = x^n(x - y) + y(x^n - y^n)$.

In Exercises 11–22, use mathematical induction to establish the given equation or inequality, where n is an arbitrary natural number. (Cf. Exs. 13–18, § 318.)

11. $1 + 2 + \cdots + n = \frac{1}{2}n(n + 1)$. (Cf. Example 1, § 317.)

12. $1^2 + 2^2 + \cdots + n^2 = \frac{1}{6}n(n + 1)(2n + 1)$. (Cf. Example 2, § 317.)

13. $1^3 + 2^3 + \cdots + n^3 = \frac{1}{4}n^2(n + 1)^2 = (1 + 2 + \cdots + n)^2$. (Cf. Ex. 7, § 318.)

14. $1^4 + 2^4 + \cdots + n^4 = \frac{1}{30}n(n + 1)(2n + 1)(3n^2 + 3n - 1)$. (Cf. Ex. 8, § 318.)

15. $1^5 + 2^5 + \cdots + n^5 = \frac{1}{12}n^2(n + 1)^2(2n^2 + 2n - 1)$. (Cf. Ex. 9, § 318.)

16. $1 + 3 + 5 + \cdots + (2n - 1) = n^2$. (Cf. Ex. 10, § 318.)

17. $2 + 5 + 8 + \cdots + (3n - 1) = \frac{1}{2}n(3n + 1)$. (Cf. Ex. 11, § 318.)

18. $1^2 + 4^2 + 7^2 + \cdots + (3n - 2)^2 = \frac{1}{2}n(6n^2 - 3n - 1)$. (Cf. Ex. 12, § 318.)

19. $\dfrac{1}{1 \cdot 2} + \dfrac{1}{2 \cdot 3} + \cdots + \dfrac{1}{n(n + 1)} = \dfrac{n}{n + 1}$.

20. $2^n > n^2$, for $n > 4$.

21. $(1 + a)^n \geq 1 + na$, if $a \geq -1$.

22. $(1 + a)^n \geq 1 + na + \frac{1}{2}n(n - 1)a^2$, if $a \geq 0$;
$(1 + a)^n \leq 1 + na + \frac{1}{2}n(n - 1)a^2$, if $-1 \leq a \leq 0$.

23. If a, b, and c are arbitrary but fixed natural numbers, prove that there must exist a natural number N such that the inequality $n > N$ implies $n^3 > a + bn + cn^2$.

24. Prove that for every natural number n the natural number $5^n - 8n^2 + 4n - 1$ is divisible by 64; that is, it can be written as a product of the form 64 m.

25. If $x_1, x_2, \cdots, x_n$ and $y_1, y_2, \cdots, y_n$ are arbitrary numbers and if $x_k \leq y_k$ for $k = 1, 2, \cdots, n$, prove that $x_1 + x_2 + \cdots + x_n \leq y_1 + y_2 + \cdots + y_n$, equality holding in the last inequality if and only if $x_k = y_k$ for every k.

26. If $x_1, x_2, \cdots, x_n$ and $y_1, y_2, \cdots, y_n$ are arbitrary numbers and if $0 \leq x_k \leq y_k$, for $k = 1, 2, \cdots, n$, prove that $x_1 x_2 \cdots x_n \leq y_1 y_2 \cdots y_n$, and find a necessary and sufficient condition for equality to hold in the last inequality.

27. Prove the following properties of min and max, defined in Example 4, § 306:

(*i*) $\max(-a_1, -a_2, \cdots, -a_n) = -\min(a_1, a_2, \cdots, a_n)$.

(*ii*) $\min(a_1 + c, \cdots, a_n + c) = \min(a_1, \cdots, a_n) + c$,
$\max(a_1 + c, \cdots, a_n + c) = \max(a_1, \cdots, a_n) + c$.

(*iii*) If $c > 0$, $\min(ca_1, \cdots, ca_n) = c \min(a_1, \cdots, a_n)$,
$\max(ca_1, \cdots, ca_n) = c \max(a_1, \cdots, a_n)$.

(iv) If $a_k \leq b_k$, for $1 \leq k \leq n$, min $(a_1, \cdots, a_n) \leq$ min $(b_1, \cdots, b_n)$, max $(a_1, \cdots, a_n) \leq$ max $(b_1, \cdots, b_n)$.

(v) min $(a_1 + b_1, \cdots, a_n + b_n) \geq$ min $(a_1, \cdots, a_n) +$ min $(b_1, \cdots, b_n)$, max $(a_1 + b_1, \cdots, a_n + b_n) \leq$ max $(a_1, \cdots, a_n) +$ max $(b_1, \cdots, b_n)$, and a strict inequality may occur in each case.

(vi) If $a_k \geq 0$ and $b_k \geq 0$ for $1 \leq k \leq n$, min $(a_1 b_1, \cdots, a_n b_n) \geq$ min $(a_1, \cdots, a_n) \cdot$ min $(b_1, \cdots, b_n)$, max $(a_1 b_1, \cdots, a_n b_n) \leq$ max $(a_1, \cdots, a_n) \cdot$ max $(b_1, \cdots, b_n)$, and a strict inequality may occur in each case.

In Exercises 28–29, use mathematical induction to establish the inequalities, where n is a natural number greater than 1.

28. $\dfrac{3}{2} - \dfrac{1}{n} + \dfrac{1}{2n^2} < \dfrac{1}{1^2} + \dfrac{1}{2^2} + \cdots + \dfrac{1}{n^2} < 2 - \dfrac{1}{n}$.

29. $\dfrac{1}{1 + 1^2} + \dfrac{1}{1 + 2^2} + \cdots + \dfrac{1}{1 + n^2} < \dfrac{5}{4} - \dfrac{1}{n}$.

30. Show that if $P(n)$ is the proposition $3 + 5 + \cdots + (2n + 1) = (n + 1)^2$, then $P(n)$ implies $P(n + 1)$ for every natural number n. Is $P(n)$ true?

31. Find the fallacy in the following "proof" that every number in the set

$$A = \left\{ -1, -\dfrac{1}{2}, -\dfrac{1}{3}, \cdots, -\dfrac{1}{n}, \cdots, \cdots, \dfrac{1}{n}, \cdots, \dfrac{1}{3}, \dfrac{1}{2}, 1 \right\}$$

is negative: In the first place, the first member of A is negative. In the second place, the successor to every negative member of A is negative. Therefore every member of A is negative.

32. Find the fallacy in the following "proof" that whenever n points $x_1, x_2, \cdots, x_n$ are given, they are all equal: $x_1 = x_2 = \cdots = x_n$: Let $P(n)$ be the proposition just stated. Then $P(1)$ is obviously true. Assuming the truth of $P(n)$ for a natural number n we wish to infer the truth of $P(n + 1)$; that is, that $x_1 = x_2 = \cdots = x_n = x_{n+1}$. From the induction assumption we have $x_1 = x_2 = \cdots = x_n$, and $x_2 = x_3 = \cdots = x_n = x_{n+1}$, and consequently $x_1 = x_2 = \cdots = x_n = x_{n+1}$, as desired. The truth of $P(n)$ for all natural numbers n now follows from the Fundamental Theorem of Mathematical Induction.

33. Prove the following variant of the Fundamental Theorem of Mathematical Induction (Theorem III, § 303): *Let N be an arbitrary but fixed natural number, and for every natural number $n \geq N$ let $P(n)$ be a proposition that is either true or false. If (i) $P(N)$ is true and (ii) whenever the proposition $P(n)$ is true for the natural number $n \geq N$ the proposition $P(n + 1)$ is also true, then $P(n)$ is true for every natural number $n \geq N$.* *Hint:* For $1 \leq n < N$ let $P(n)$ be any true proposition. (Cf. Ex. 12, § 506.)

309. CARTESIAN PRODUCTS, AND RELATIONS

If A and B are any two nonempty sets, the **Cartesian product of A and B**, denoted by $A \times B$, is the set of all ordered pairs† (x, y), where $x \in A$ and

---

† No attempt will be made in this book to define the concept of *ordered pair* in more primitive terms. (Cf. Robert R. Stoll, *Sets, Logic and Axiomatic Theories* (San Francisco,

$y \in B$. Any subset ρ of $A \times B$ is called a **relation from A to B**. The notation $x \,\rho\, y$ means that the ordered pair (x, y) belongs to the subset ρ of $A \times B$: $(x, y) \in \rho$. The set of all x in A such that $x \,\rho\, y$ for *some* y in B is called the **domain**, or **domain of definition**, of ρ. The set of all y in B such that $x \,\rho\, y$ for *some* x in A is called the **range**, or **range of values**, of ρ. The Cartesian product $\mathscr{R} \times \mathscr{R}$ of the real number system and itself is called the **Cartesian plane**, and is written $\mathscr{R}^2$.

Example 1. Let A be the set of all women x living at a particular time t, let B be the set of all men y living at time t, and let ρ be the set of all pairs (x, y) where $x \in A$, $y \in B$, and x and y are married to each other. The relation ρ in this case is that of "having as a husband." The domain of ρ is the set of all wives (at time t), and the range of ρ is the set of all husbands. (For a second look at this example, cf. Example 1, § 310.)

Example 2. Let A and B be the set of all people who have ever lived, and let ρ be the set of all pairs (x, y) where $x \in A$ and $y \in B$, and y is the mother of x. The relation ρ in this case is that of "having as a mother." The domain of ρ is A itself, and the range of ρ is the set of all mothers. (For a second look at this example, cf. Example 2, § 310.)

Several different types of relations are important in mathematics. In Chapter 6, for instance, *equivalence relations* receive principal attention. In the present chapter we shall be concerned only with the type of relation known as a *function*, defined in the following section.

310. FUNCTIONS

Definition. *Let A and B be any two nonempty sets. Then a **function**[†] from A to B is a relation f from A to B such that no two distinct members of f have the same first coordinate: if (x, y_1) and (x, y_2) are members of the set f, where $x \in A$, $y_1 \in B$, and $y_2 \in B$, then $y_1 = y_2$. The function f is a **function on A into B** if and only if f is a function from A to B with domain of definition equal to A; that is, every member x of A is the first coordinate of some pair (x, y) belonging to f. If x denotes an arbitrary member of the domain of a function f, then f is called a function of the **variable** x. A function f is a **one-to-one correspondence** between the members of A and the members of B if and only if f is a function on A into B with range equal to B such that no*

W. H. Freeman and Company, 1961), p. 26, for further discussion.) However, it should be noted that *equality* between two ordered pairs (x, y) and (u, v), written $(x, y) = (u, v)$, means that $x = u$ and $y = v$, and that if $x \neq y$, then the ordered pairs (x, y) and (y, x) are distinct: $(x, y) \neq (y, x)$. In the ordered pair (x, y), x and y are called **coordinates**, x the **first coordinate** and y the **second coordinate**.

† The present definition prescribes what is sometimes called a *single-valued* function. In many contexts it is valuable to admit *multiple-valued* functions to the general family of functions (for example, this is true in certain parts of both real and complex variable theory). For this more general interpretation of function, the concepts of *relation* and *function* are identical. In this book we shall restrict ourselves to the definition given above.

*two distinct members of f have the same second coordinate: if (x_1, y) and (x_2, y) are members of the set f, where $x_1 \in A$, $x_2 \in A$, and $y \in B$, then $x_1 = x_2$. The **values** of a function are the members of its range of values. A **constant** **function** is a function whose range consists of one point. If $\mathcal{R}$ denotes the real number system, if f is a function from A to B, and if $A \subset \mathcal{R}$, then f is a function of a **real variable**; if $B \subset \mathcal{R}$, then f is **real-valued**.*

Example 1. In Example 1, § 309, if the world were devoid of polyandry at time t, the relation ρ would be a function, since every married woman would have one and only one husband. If, in addition, the world were strictly monogamous (at most one mate) at time t, then ρ would be a one-to-one correspondence between wives and husbands.

Example 2. In Example 2, § 309, the relation ρ is a function. If the domain were restricted to first-born children the relation would be a one-to-one correspondence.

Example 3. The set of all ordered pairs of real numbers of the form (x, x^2) is a real-valued function of a real variable with domain of definition $\mathcal{R}$. The set of all ordered pairs of real numbers of the form (x, x^2), where $1 \leq x < 7$, is a different real-valued function of a real variable with domain of definition the set of numbers x satisfying the inequalities $1 \leq x < 7$.

Discussion and Notation. The preceding definition guarantees that if f is a function on A into B, then every member x of A determines *exactly one* member y of B, which is then said to **correspond** to x. This correspondence is written: $y = f(x)$. Conversely, if some formula or rule exists whereby to each point x of A there corresponds exactly one point y of B, then the resulting set of all ordered pairs (x, y) such that y corresponds to x constitutes a function on A into B as defined above. The statement that f is a function on A into B is expressed symbolically either in the form $f: A \to B$ or in the form $A \xrightarrow{f} B$. Thus, either $f: \mathcal{R} \to \mathcal{R}$ or $\mathcal{R} \xrightarrow{f} \mathcal{R}$ states that f is a real-valued function of a real variable and that $f(x)$ is defined for every real number x. It is frequently convenient, especially when discussing specific functions, to abbreviate what would otherwise be a rather elaborate and awkward sentence structure, by permitting the letter x to play a direct role as follows: Such a phrase as "a function $f(x)$, where $1 < x < 2$" should be interpreted as an abbreviation for "a function f whose domain of definition is the set of all real numbers x such that $1 < x < 2$," and the statement "consider the function $x^2 + 2x$, $x \in \mathcal{R}$" is a compression of the longer statement "consider the function f on $\mathcal{R}$ into $\mathcal{R}$ defined by the equation $f(x) = x^2 + 2x$." In this last illustration the fact that the expression $x^2 + 2x$ is used to represent a function whereas, for any particular x, $x^2 + 2x$ is a number should cause no more confusion—if the context is clear—than saying "this is John Smith" instead of "this is a man whose name is John Smith." A one-to-one correspondence is often denoted by means of a double-headed arrow $\leftrightarrow$, as illustrated in Examples 5 and 6, below.

Note 1.   If a **vertical line** in the Cartesian plane $\mathscr{R} \times \mathscr{R}$ (cf. § 309) is defined to be the set of all "points" or ordered pairs (x, y) whose first coordinate is some fixed real number x, a real-valued function of a real variable can be pictured as any subset of the Cartesian plane no two points of which lie on the same vertical line.   For readers familiar with analytic geometry, this set of points is what is commonly called the *graph of a function*.   The essential idea of the preceding definition is to identify the fundamental concepts of *function* and *graph of a function*, without destroying the idea of correspondence associated with a function or that of a geometrical figure associated with a graph of a function.

Note 2.   *Equality* between functions f and g simply means that as sets of ordered pairs they are identical sets: f and g have identical domains of definition and identical ranges, and for every x of their common domain $f(x) = g(x)$.   Equality between functions, then, has nothing to do with the particular manner in which they may be defined.   For example, the functions $f(x) = x^2 + 2x$ and $g(x) = x(x + 2)$ are the same, although one is defined as a sum and the other as a product.

Note 3.   If f and g are real-valued functions with a common domain A and if c is any real number, the functions $f \pm g$, fg, and cf, all with domain A, are defined:

$$(1) \qquad \begin{cases} h = f \pm g & : h(x) = f(x) \pm g(x) \\ h = fg & : h(x) = f(x)g(x) \\ h = cf & : h(x) = cf(x). \end{cases}$$

The function f/g is defined, with domain consisting of all points x of A such that $g(x) \neq 0$:

$$(2) \qquad\qquad h = \frac{f}{g} : h(x) = \frac{f(x)}{g(x)}.$$

Example 4.   The function $n!$, called the **factorial function.** whose domain of definition consists of 0 and the natural numbers, is defined as follows: $0! = 1$; if n is a natural number, $n!$ is the product of the n natural numbers $1, 2, \cdots, n$.   Thus, $0! = 1$, $1! = 1$, $2! = 2$, $3! = 6$, $4! = 24$, $5! = 120$, $6! = 720$, $7! = 5040$, $8! = 40{,}320$, $9! = 362{,}880$, and $10! = 3{,}628{,}800$.

Example 5.   The set of all ordered pairs $(n, 2^n)$, where $n \in \mathscr{N}$, is a function on $\mathscr{N}$ into $\mathscr{N}$.   It is a one-to-one correspondence between the members of its domain and those of its range (cf. Ex. 6, § 308), alternatively denoted $n \leftrightarrow 2^n$.

Example 6.   The set of all ordered pairs $(n, 2n - 1)$, where $n = 1, 2, 3$, or 4, is a one-to-one correspondence between the points 1, 2, 3, 4 and the points 1, 3, 5, 7, also denoted $n \leftrightarrow 2n - 1$.

Example 7.   The set of all ordered pairs (x^2, x), $x \in \mathscr{R}$ is not a (single-valued) function, since $(1, 1)$ and $(1, -1)$ are distinct members of this set possessing the same first coordinate.

Example 8.   Show that any set of n members of an ordered field $\mathscr{G}$ can be arranged in increasing order.   That is, if $x_1, x_2, \cdots, x_n$ are members of $\mathscr{G}$, then there exists a one-to-one correspondence $k \leftrightarrow m_k$, $k = 1, 2, \cdots, n$, between the first n natural numbers $1, 2, \cdots, n$ and the same first n natural numbers $1, 2, \cdots, n$ such that $x_{m_1} \leqq x_{m_2} \leqq \cdots \leqq x_{m_n}$; that is, if $y_k = x_{m_k}$, $k = 1, 2, \cdots, n$, then $y_1 \leqq y_2 \leqq \cdots \leqq y_n$.

Solution.   Let $P(n)$ be the proposition just stated.   Then $P(1)$ is trivially true.

Assuming now the truth of $P(n)$ for a particular natural number n, we wish to infer that of $P(n + 1)$. Accordingly, let $x_1, x_2, \cdots, x_n, x_{n+1}$ be any $n + 1$ members of $\mathscr{G}$. By Example 4, § 306, there exists a natural number j, where $1 \leqq j \leqq n + 1$, such that $x_j = \max (x_1, x_2, \cdots, x_{n+1})$. Define $z_1, z_2, \cdots, z_n$ as follows: If $j = 1$, let $z_k \equiv x_{k+1}$, $k = 1, 2, \cdots, n$; if $j = n + 1$, let $z_k \equiv x_k$, $k = 1, 2, \cdots, n$; if $1 < j < n + 1$, let $z_k \equiv x_k$ for $k = 1, 2, \cdots, j - 1$ and let $z_k \equiv x_{k+1}$ for $k = j, j + 1,$ $\cdots, n$. In other words, the z's are simply the x's that remain after deletion of x_j from $\{x_1, x_2, \cdots, x_{n+1}\}$. By the induction assumption, the n numbers $z_1, z_2, \cdots,$ z_n—or, equivalently, the n numbers $x_2, x_3, \cdots, x_{n+1}$, or $x_1, x_2, \cdots, x_n$, or $x_1, x_2,$ $\cdots, x_{j-1}, x_{j+1}, \cdots, x_{n+1}$ (as the case may be)—can be arranged in increasing order, denoted: $y_1 \leqq y_2 \leqq \cdots \leqq y_n$. If $y_{n+1} \equiv x_j$, then the entire set of numbers $x_1, x_2, \cdots, x_{n+1}$, when relabeled $y_1, y_2, \cdots, y_{n+1}$, becomes rearranged in increasing order: $y_1 \leqq y_2 \leqq \cdots \leqq y_{n+1}$. An application of the Fundamental Theorem of Mathematical Induction completes the proof.

311. A NATURAL ISOMORPHISM

It will be shown in Chapters 5, 6, 7, and 9 that it is possible to have many different kinds of fields (for example, both finite and infinite), and that there is some distinctive variety among ordered fields as well. In the face of this complexity a natural question to ask is "In what way are two natural number systems, one from one ordered field and one from another ordered field, related?" It is conceivable, at least, that one natural number system might have some algebraic property not holding in some other natural number system. It is the purpose of this section to show that such is not the case. We shall prove that, in a manner to be made precise, *any two natural number systems are abstractly identical.* In other words, when we make use of the properties of the natural numbers it is immaterial from what ordered field they arise.

Definition. *Let A be a subset of an ordered field $\mathscr{G}$ and let A be closed with respect to addition and multiplication (cf. the Note, § 202), and let A' be a subset of an ordered field $\mathscr{G}'$ similarly closed with respect to operations in $\mathscr{G}'$. Then A and A' are* **isomorphic** *if and only if there exists a one-to-one correspondence between their members preserving the operations of addition and multiplication, as well as preserving the order relation; that is, if and only if there exists a one-to-one correspondence denoted $a \leftrightarrow a'$, where $a \in A$ and $a' \in A'$, such that the two correspondences*

$$a_1 \leftrightarrow a_1' \quad and \quad a_2 \leftrightarrow a_2'$$

imply the two correspondences

$$a_1 + a_2 \leftrightarrow a_1' + a_2' \quad and \quad a_1 a_2 \leftrightarrow a_1' a_2',$$

or, in another form,

$$(a_1 + a_2)' = a_1' + a_2' \quad and \quad (a_1 a_2)' = a_1' a_2',$$

and, furthermore,

$$a_1 < a_2 \quad if \text{ and } only \text{ if } \quad a_1' < a_2'.$$

*Under these conditions the correspondence $a \leftrightarrow a'$ is called an **isomorphism**.
Such an isomorphism is said to **preserve** addition, multiplication, and order,
or to be **addition-preserving**, **multiplication-preserving**, and **order-
preserving**.*

Theorem.   *Any two natural number systems are isomorphic.   That is, if
$\mathcal{N}$ is the system of natural numbers of an ordered field $\mathcal{G}$ and if $\mathcal{N}'$ is the
system of natural numbers of an ordered field $\mathcal{G}'$, then $\mathcal{N}$ and $\mathcal{N}'$ are isomorphic.*

Proof.   In setting up a one-to-one correspondence between the natural
numbers n of $\mathcal{N}$ and the natural numbers n' of $\mathcal{N}'$ we start by letting the
unity elements, 1 of $\mathcal{N}$ and $1'$ of $\mathcal{N}'$, correspond: $1 \leftrightarrow 1'$.   We next extend
the correspondence to all natural numbers by declaring that whenever
$n \leftrightarrow n'$, then $n + 1 \leftrightarrow n' + 1'$.   By the Fundamental Theorem of Mathematical Induction this correspondence is defined for every natural number n
of $\mathcal{N}$ and for every natural number n' of $\mathcal{N}'$, and by the same principle the
correspondence is one-to-one.   We wish next to show that the correspondence
$\leftrightarrow$ is an isomorphism with respect to addition; that is, that if $m \leftrightarrow m'$ and
$n \leftrightarrow n'$, then $m + n \leftrightarrow m' + n'$.   This is again done by induction: For a
fixed natural number m of $\mathcal{N}$, $m + 1 \leftrightarrow m' + 1'$ by definition, and if
$m + n \leftrightarrow m' + n'$, then $m + (n + 1) = (m + n) + 1 \leftrightarrow (m + n)' + 1' =
(m' + n') + 1' = m' + (n' + 1') = m' + (n + 1)'$.   It is proved similarly
that $\leftrightarrow$ is an isomorphism with respect to multiplication, for natural numbers:
$(mn)' = m'n'$.   In detail: For a fixed natural number m of $\mathcal{N}$, $(m \cdot 1)' =
m' \cdot 1'$, and if $(mn)' = m'n'$, then $(m(n + 1))' = (mn + m)' = (mn)' + m' =
m'n' + m' \cdot 1' = m'(n' + 1') = m'(n + 1)'$, where the isomorphism with
respect to addition, already established, is occasionally invoked.   Finally,
the correspondence is order-preserving since, if $m > n$ there exists a natural
number p such that $m = n + p$.   Therefore, $m' = n' + p'$, and $m' > n'$.
Conversely, we see by reversing these steps that $m' > n'$ implies $m > n$.

Corollary.   *The correspondence established between the members of $\mathcal{N}$ and
$\mathcal{N}'$ in the proof of the preceding theorem is an isomorphism with respect to
subtraction when it is defined; that is, if m and n are natural numbers in $\mathcal{N}$,
if $m > n$, and if $m \leftrightarrow m'$ and $n \leftrightarrow n'$, then $m - n \leftrightarrow m' - n'$, or equivalently,
$(m - n)' = m' - n'$.*

Proof.   Since $m - n$ is a natural number in $\mathcal{N}$ such that $n + (m - n) = m$,
the isomorphism with respect to addition, already established, gives the equation $n' + (m - n)' = m'$.   Solving for $(m - n)'$ gives the desired equality.

It is left as an exercise for the reader to show that the correspondence
discussed above is an isomorphism with respect to division whenever division
is defined.

312. FINITE AND INFINITE SETS

Definition I, § 305, contains the essential ideas of what is meant by a
finite set: a set of n distinct points $x_1, x_2, \cdots, x_n$, for some natural number n.

The following definition is in essence a restatement of this, but permits the inclusion of the empty set.

Definition. *A set A is **finite** if and only if it is empty or its members can be put into a one-to-one correspondence (cf. § 310) with the natural numbers 1, 2, $\cdots$, n, for some n; that is—for a nonempty set—if and only if there exists a set of ordered pairs (k, x) where k is a natural number with $1 \leq k \leq n$ and x is a member of A, such that to each k for which $1 \leq k \leq n$ there corresponds exactly one ordered pair (k, x) having this particular k as its first coordinate, and to each point x of A there corresponds exactly one ordered pair (k, x) having this particular x as its second coordinate. A set is **infinite** if and only if it is not finite.*

Our first theorem concerning finite and infinite sets is intuitively obvious, but nevertheless requires proof. As might be expected, the proof is by mathematical induction. The letters m, n, r, and j denote natural numbers.

Theorem I. *Any subset of a finite set is finite. Equivalently, any set that contains an infinite subset is infinite. If $A = \{a_1, a_2, \cdots, a_m\}$, where the a's are distinct, if $B = \{b_1, b_2, \cdots, b_n\}$, where the b's are distinct, and if A is a subset of B, then $m \leq n$. If, in addition, A is a proper subset of B, then $m < n$.*

Proof. Clearly, the empty set has only itself as subset, and this is finite. Let $P(n)$ be the proposition that if B is the set of distinct points $b_1, b_2, \cdots, b_n$ and if A is a subset of B then A is finite, and furthermore that if A consists of the m distinct points $a_1, a_2, \cdots, a_m$ then $m \leq n$, with the strict inequality $m < n$ holding in case A is a proper subset of B. The proposition $P(1)$ holds trivially. Assuming now that $P(n)$ is true for a particular natural number n, we shall show that $P(n + 1)$ is also true. Accordingly, let B consist of $n + 1$ distinct points: $B = \{b_1, b_2, \cdots, b_n, b_{n+1}\}$, and let A be a subset of B. There are two cases to consider according as b_{n+1} is or is not a member of A. If the point b_{n+1} is *not* a member of A, then A is a *proper* subset of B and, furthermore, A is a subset of $\{b_1, b_2, \cdots, b_n\}$. By the induction assumption that $P(n)$ is true, A is therefore a finite set, and if $A = \{a_1, a_2, \cdots, a_m\}$ then $m \leq n < n + 1$, as desired. On the other hand, if b_{n+1} is a member of A, $b_{n+1} = a_r$ where $1 \leq r \leq m$. Consider the set $C \equiv A - \{a_r\}$ obtained by deleting the point $b_{n+1} = a_r$ from the set A. Since $C \subset \{b_1, b_2, \cdots, b_n\}$, C is a finite set by the induction assumption. Let the point c_j be defined to be a_j for all j such that $1 \leq j < r$ and let c_j be defined to be a_{j+1} for all j such that $r \leq j \leq m - 1$. We thus have C represented:

(1)
$$C = \begin{cases} \{a_1, a_2, \cdots, a_{r-1}, a_{r+1}, \cdots, a_m\}, \\ \{c_1, c_2, \cdots, c_{r-1}, c_r, \cdots, c_{m-1}\}. \end{cases}$$

(If $r = m$, the upper line of (1) reads $\{a_1, a_2, \cdots, a_{m-1}\}$ without a gap; if $r = 1$, the upper line of (1) reads $\{a_2, a_3, \cdots, a_m\}$, also without a gap.)

Notice that since b_{n+1} has been deleted from the set A in forming C, C is a subset of $\{b_1, b_2, \cdots, b_n\}$ and, furthermore, that C is a proper subset of $b_1, b_2, \cdots, b_n$ if and only if A is a proper subset of B. We now use the induction assumption for $P(n)$, concluding that $m - 1 \leqq n$, or $m \leqq n + 1$, in every case, and that if A is a proper subset of B, $m - 1 < n$, or $m < n + 1$. With final appeal to the Fundamental Theorem of Mathematical Induction, the proof is complete.

As a corollary, we have the theorem:

Theorem II. *The number of points in a finite set is uniquely determined. That is (except for the trivial case of the empty set), if* $\{a_1, a_2, \cdots, a_m\} = \{b_1, b_2, \cdots, b_n\}$, *if the a's are distinct, and if the b's are distinct, then* $m = n$.

Proof. Since the set of a's is a subset of the set of b's, $m \leqq n$. Since the set of b's is a subset of the set of a's, $n \leqq m$. Therefore $m = n$.

As a further corollary we have:

Theorem III. *If the points of a set A can be put into a one-to-one correspondence with the points of a finite set B, then A is a finite set having the same number of points as B.*

Proof. Since the points of B are in one-to-one correspondence with the natural numbers $1, 2, \cdots, n$, for some n (except for the trivial case of the empty set), the points of A are also in one-to-one correspondence with this same set of natural numbers and hence can be written $a_1, a_2, \cdots, a_n$.

Another result that can be inferred from these facts is:

Theorem IV. *The points of a finite set cannot be put into a one-to-one correspondence with those of a proper subset.*

Proof. In Theorem III, if A denotes a proper subset of a finite set B, assuming that a one-to-one correspondence between their members is possible, a contradiction to the final statement of Theorem I is obtained.

NOTE. Theorem IV shows that *if* the points of a set can be put into a one-to-one correspondence with those of a proper subset, then the set must be infinite. It does *not* show that the points of any infinite set can be put into such correspondence. The German mathematician R. Dedekind (1831–1916) *defined* an infinite set to be one whose members can be put into a one-to-one correspondence with those of a proper subset. The reader interested in reading more about infinite sets will find stimulating material and further references by consulting the topics *Dedekind* and *infinite* in the index of James R. Newman, *The World of Mathematics* (New York, Simon and Schuster, 1956).

Theorem IV provides us with our first proof that the set of natural numbers is infinite (cf. Theorem VII):

Theorem V. *The natural numbers of any ordered field constitute an infinite set. Consequently every ordered field is infinite.*

Proof. The correspondence $n \leftrightarrow n + 1$ is a one-to-one correspondence between the natural numbers n and the natural numbers m that satisfy the inequality $m > 1$. Specifically, corresponding to any natural number n is the natural number $n + 1$ that satisfies the inequality $n + 1 > 1$ (since $n > 0$). On the other hand, if m is a natural number such that $m > 1$, then (by V, § 304) $m - 1$ is a natural number to which m corresponds by the process of adding 1. Finally, since the set of natural numbers greater than 1 is a proper subset of the set of *all* natural numbers (including 1), it is impossible for the set of natural numbers to be finite. Finally, by Theorem I, since every ordered field contains a system of natural numbers, which is infinite, the ordered field must be infinite as well.

At this point we pause to record an important fact about finite sets of real numbers (or more generally, of any ordered field), which is an immediate consequence of Example 4, § 306:

Theorem VI. *Any nonempty finite set in an ordered field contains a least member and a greatest member.*

We conclude this section with a theorem which, in conjunction with Theorem VI, provides us with a second proof of the infinitude of the natural numbers and of the members of any ordered field.

Theorem VII. *There is no greatest natural number. There is no greatest member of any ordered field.*

Proof. Corresponding to any natural number n, or member x of an ordered field, there exists a natural number $n + 1$ that is greater than n, or a member $x + 1$ that is greater than x.

313. SEQUENCES

An informal definition of a sequence can be framed as follows: Start with the first term, a_1; then proceed to the second term, a_2; then to the third, a_3; etc. If the sequence terminates, it is a *finite sequence;* otherwise, it is an *infinite sequence.* A more formal definition follows:

Definition I. *A finite sequence is a function whose domain of definition is the finite set of natural numbers $\{1, 2, \cdots, m\}$, for some natural number m; it is written $a_1, a_2, \cdots, a_m$, or $\{a_n\}, n = 1, 2, \cdots, m$. An infinite sequence is a function whose domain of definition is the set of all natural numbers and is written $a_1, a_2, \cdots, a_n, \cdots$, or $\{a_n\}$. The values a_n of a sequence are also called the* **terms** *of the sequence.*

Examples 1. The following are examples of finite sequences of natural numbers:

(1) 1;
(2) 6, 6;
(3) 1, 5, 3;
(4) 7, 5 3, 1;
(5) 5, 5, 4, 4, 2.

Examples 2. The following are examples of infinite sequences of real numbers:

(6) $1, 2, 3, \cdots, n, \cdots$;

(7) $1, \dfrac{1}{2}, \dfrac{1}{3}, \cdots, \dfrac{1}{n}, \cdots$;

(8) $1, 1, 1, \cdots, 1, \cdots$;
(9) $0, 1, 0, 1, \cdots, \frac{1}{2}(1 + (-1)^n), \cdots$;
(10) $1, 0, 2, 0, \cdots, \frac{1}{4}(n + 1)(1 + (-1)^{n+1}), \cdots$;
(11) $1, 1, \frac{1}{2}, \frac{1}{2}, \cdots, a_n, \cdots$, where $a_n = \begin{cases} 2/(n + 1) \text{ if } n \text{ is odd,} \\ 2/n \text{ if } n \text{ is even.} \end{cases}$

314. ARITHMETIC AND GEOMETRIC SEQUENCES

In this section we shall define two particular types of sequence, and obtain formulas for the sums of the first n terms for each.   The word *number* is used to designate a member of an arbitrary ordered field.

Definition I. *A finite or infinite sequence $\{a_n\}$ of real numbers (or, more generally, of members of an arbitrary ordered field) is an **arithmetic sequence** or **arithmetic progression** if and only if there exist numbers a and d such that $a_n = a + (n - 1)d$ for every term of the sequence.   Since $a_n - a_{n-1} = d$ for every term a_n after the first, d is called the **common difference** of the sequence.*

Examples 1. Each of the following is an arithmetic sequence, with common difference as specified:

(1) 1; $d = 17$;
(2) $1, 3, 5, 7$; $d = 2$;
(3) $6, 6, 6, 6, \cdots, 6, \cdots$; $d = 0$;
(4) $0, -1, -2, -3, \cdots, -n + 1, \cdots$; $d = -1$.

Definition II. *A finite or infinite sequence $\{a_n\}$ of real numbers (or, more generally, of members of an arbitrary ordered field) is a **geometric sequence or geometric progression** if and only if there exist numbers a and r such that $a_n = ar^{n-1}$ for every term of the sequence (where $r^0 = 1$ for every number r, including $r = 0$—cf. Ex. 4, § 308).   Since, if $ar \neq 0$, $a_n/a_{n-1} = r$ for every term a_n after the first, r is called the **common ratio** of the sequence.*

Examples 2. Each of the following is a geometric sequence, with common ratio as specified:

(5) 1; $r = 29$;
(6) $3, 6, 12, 24, 48$; $r = 2$;

(7) $0, 0, 0, \cdots, 0, \cdots ; \; r = 13;$

(8) $3, 0, 0, 0, \cdots, 0, \cdots ; \; r = 0;$

(9) $27, 18, 12, 8, \cdots, 3^3 \cdot 2^{n-1}/3^{n-1}, \cdots, r = 2/3.$

Theorem I. *If $a_1, a_2, \cdots, a_n$ are the first n terms of an arithmetic sequence, where $a_n = a + (n - 1)d$ for $n = 1, 2, \cdots$, and if $S_n \equiv a_1 + a_2 + \cdots + a_n = a + (a + d) + \cdots + (a + (n - 1)d)$, then*

(10) $$S_n = na + \tfrac{1}{2}(n - 1)nd = \tfrac{1}{2}n(a_1 + a_n).$$

Proof. The second equality of (10) is a simple matter of substitution for a_1 and a_n. Let $P(m)$ be the proposition that $S_m = ma + \tfrac{1}{2}(m - 1)md$ if a_m is a term of the sequence, and let $P(m)$ be any true statement in case the sequence is finite and has fewer than m terms. $P(1)$ is trivially true. Assuming that the sequence has at least $m + 1$ terms and that $P(m)$ is true, we wish to show that

(11) $$S_{m+1} = (m + 1)a + \tfrac{1}{2}m(m + 1)d.$$

The equality (11) follows from the induction hypothesis and the fact that $S_{m+1} = S_m + a_{m+1}$: $S_{m+1} = [ma + \tfrac{1}{2}(m - 1)md] + [a + md] = (m + 1)a + \tfrac{1}{2}m(m + 1)d$. By the Fundamental Theorem of Mathematical Induction the proof is complete.

Examples 3. For the sequences of Examples 1, and the specified value of n, formula (10) gives:

(1) $n = 1: \; 1 = 1 \cdot 1 + \tfrac{1}{2} \cdot 0 \cdot 1 \cdot 17 = \tfrac{1}{2} \cdot 1(1 + 1);$

(2) $n = 4: \; 16 = 4 \cdot 1 + \tfrac{1}{2} \cdot 3 \cdot 4 \cdot 2 = \tfrac{1}{2} \cdot 4(1 + 7);$

(3) $n = n: \; 6n = n \cdot 6 + \tfrac{1}{2}(n - 1)n \cdot 0 = \tfrac{1}{2}n(6 + 6);$

(4) $n = n: \; \tfrac{1}{2}(n - n^2) = n \cdot 0 + \tfrac{1}{2}(n - 1)n(-1) = \tfrac{1}{2}n(0 - n + 1).$

Theorem II. *If $a_1, a_2, \cdots, a_n$ are the first n terms of a geometric progression, where $a_n = ar^{n-1}$ for $n = 1, 2, \cdots$, and if $S_n \equiv a_1 + a_2 + \cdots + a_n = a + ar + ar^2 + \cdots + ar^{n-1}$, then*

(12) $$S_n = \begin{cases} na, & \text{if } r = 1, \\ \dfrac{a(1 - r^n)}{1 - r} = \dfrac{a_1 - ra_n}{1 - r}, & \text{if } r \neq 1. \end{cases}$$

Proof. If $r = 1$, the sequence is an arithmetic sequence with $d = 0$ and, by (10), $S_n = na$. Assume now that $r \neq 1$. Then the equality on the second line of (12) is a simple matter of substitution for $a_1 = a$ and $a_n = ar^{n-1}$. Let $P(m)$ be the proposition that $S_m = a(1 - r^m)/(1 - r)$ if a_m is a term of the given sequence, and let $P(m)$ be any true statement in case the sequence is finite and has fewer than m terms. $P(1)$ is trivially true. Assuming that the sequence has at least $m + 1$ terms and that $P(m)$ is true, we wish to show that $S_{m+1} = a(1 - r^{m+1})/(1 - r)$. This follows from the induction hypothesis and the fact that $S_{m+1} = S_m + a_{m+1}$:

$$S_{m+1} = \frac{a(1 - r^m)}{1 - r} + \frac{ar^m(1 - r)}{1 - r} = \frac{a - ar^m + ar^m - ar^{m+1}}{1 - r} = \frac{a(1 - r^{m+1})}{1 - r}$$

By the Fundamental Theorem of Mathematical Induction the proof is complete.

Examples 4. For the sequences of Examples 2, and the specified value of n, formula (12) gives

(5) $n = 1$: $1 = \dfrac{1(1 - 29)}{1 - 29} = \dfrac{1 - 29 \cdot 1}{1 - 29}$;

(6) $n = 5$: $93 = \dfrac{3(1 - 32)}{1 - 2} = \dfrac{3 - 96}{1 - 2}$;

(7) $n = n$: $0 = \dfrac{0(1 - 13^n)}{1 - 13} = \dfrac{0 - 13 \cdot 0}{1 - 13}$;

(8) $n = n$: $3 = \dfrac{3(1 - 0)}{1 - 0} = \dfrac{3 - 0 \cdot a_n}{1 - 0}$;

(9) $n = 4$: $65 = \dfrac{27(1 - \frac{16}{81})}{1 - \frac{2}{3}} = \dfrac{27 - \frac{2}{3} \cdot 8}{1 - \frac{2}{3}}$.

315. SEQUENCES DEFINED INDUCTIVELY

It sometimes happens that the first few terms of a sequence are prescribed individually and that the remaining terms are given not by an individual formula, but by a rule that relates these values to those of preceding terms. In such a case the terms of the sequence are said to be defined **inductively**, or **recursively**, or **by recursion**. A rule that relates the general term to its predecessors is called a **recursion** or **recursive formula**.

Theorem. *Let $\{a_n\}$ be an infinite sequence (the nature of whose terms is immaterial) and let k be a natural number. If the values of $a_1, a_2, \cdots, a_k$ are prescribed, and if for every natural number n such that $n \geq k$ the value of a_{n+1} is uniquely determined by the values of the terms a_m for $1 \leq m \leq n$, then the entire infinite sequence $\{a_n\}$ is well defined; that is, a_n is uniquely determined for every natural number n.*

Proof. If $P(n)$ is the proposition that a_m is uniquely defined for $m = 1, 2, \cdots, n$, then $P(n)$ is true by assumption for $n = 1, 2, \cdots, k$. Now let n be greater than or equal to k and assume that $P(n)$ is true. Then $P(n + 1)$ must also be true since the value of a_{n+1} is uniquely determined by quantities assumed to be well defined. By the Fundamental Theorem of Mathematical Induction the theorem is established.

Example. The **Fibonacci sequence** of number theory, named after the twelfth and thirteenth century Italian mathematician Leonardo Fibonacci, is defined as follows:

(1)
$$a_n = \begin{cases} 1, \text{ if } n = 1 \text{ or } n = 2, \\ a_{n-2} + a_{n-1}, \text{ if } n > 2. \end{cases}$$

The first twelve terms are 1, 1, 2, 3, 5, 8, 13, 21, 34, 55, 89, and 144. For a single formula for a_n see Ex. 19, § 1010. For more information on Fibonacci and his sequence, see H. Eves, *An Introduction to the History of Mathematics* (New York, Holt, Rinehart and Winston, 1960), especially p. 228, and N. N. Vorob'ev, *Fibonacci Numbers*, in the Blaisdell Scientific Paperbacks series of popular lectures in mathematics translated from the Russian (New York, Blaisdell Publishing Co., 1961). (Also cf. Ex. 31, § 318, Ex. 18, § 406.)

316. DECREASING SEQUENCES OF NATURAL NUMBERS

In the present section we shall be interested in only one type of sequence (cf. § 1204 for further discussion of sequences).

Definition. *A finite or infinite sequence* $\{a_n\}$ *of real numbers (or, more generally, of members of an ordered field $\mathscr{G}$) is* **decreasing**† *if and only if* $a_n \geqq a_{n+1}$ *for every natural number n for which* a_n *and* a_{n+1} *exist. A sequence* $\{a_n\}$ *is* **strictly decreasing** *if and only if* $a_n > a_{n+1}$ *for every natural number n for which* a_n *and* a_{n+1} *exist.*

Examples. Of the finite sequence of Examples 1, § 313, the following are decreasing: (1), (2), (4), (5); the following are strictly decreasing: (1), (4). Of the infinite sequences of Examples 2, § 313, (7), (8), and (11) are decreasing, but only (7) is strictly decreasing.

For purposes of establishing the theorem that follows, we first prove the lemma:

Lemma. *If* $a_1, a_2, \cdots, a_{k+1}$ *is a strictly decreasing sequence of* $k + 1$ *natural numbers, then* $a_1 > k$.

Proof. By Example 1, § 306, if $a_1 \leqq k$, then $a_n \leqq k$ for $n = 1, 2, \cdots,$ $k + 1$, and, furthermore, the set $A \equiv \{a_1, a_2, \cdots, a_{k+1}\}$ is a finite set of $k + 1$ distinct points. However, if $a_n \leqq k$ for $n = 1, 2, \cdots, k + 1$, then A is a subset of the set $\{1, 2, \cdots, k\}$ of natural numbers and, by Theorem I, § 312, $k + 1 \leqq k$. (Contradiction.)

The following theorem establishes the "intuitively obvious" fact that however "far out" in the natural number system one may start, any "descent" must be finite.

---

† A decreasing sequence is also called a **monotonically decreasing sequence,** or a **non-increasing sequence.**

Theorem. *There does not exist an infinite strictly decreasing sequence of natural numbers. Equivalently, any strictly decreasing sequence of natural numbers is finite.*

Proof. Assume that $\{a_n\}$ is an infinite strictly decreasing sequence of natural numbers, and let $k \equiv a_1$. Then $a_1, a_2, \cdots, a_{k+1}$ is a finite strictly decreasing sequence of natural numbers, and by the Lemma, $k = a_1 > k$. (Contradiction.)

317. SIGMA SUMMATION NOTATION

Assume that to 0 and every natural number n there corresponds a uniquely determined member of the real number system $\mathscr{R}$ (or, more generally, a member of an arbitrary field $\mathscr{F}$), denoted $f(n)$ (that is, f is a function on the extended natural number system into $\mathscr{R}$—cf. Ex. 4, § 308). The sigma summation notation is defined:

$$(1) \qquad \sum_{k=m}^{n} f(k) = f(m) + f(m + 1) + \cdots + f(n),$$

where m is either 0 or a natural number, n is either 0 or a natural number, and $m \leq n$. (The sum (1) always exists since it is the sum of a finite number of terms.)

Properties of sigma summation notation are contained in the theorem:

Theorem I.
(i) *In (1), k is a* ***dummy variable:*** $\sum_{k=m}^{n} f(k) = \sum_{i=m}^{n} f(i)$.

(ii) Σ *is* ***additive;*** *that is, if f and g are any two functions on the extended natural number system into $\mathscr{R}$, then:*

$$(2) \qquad \sum_{k=m}^{n} [f(k) + g(k)] = \sum_{k=m}^{n} f(k) + \sum_{k=m}^{n} g(k).$$

(iii) Σ *is* ***homogeneous;*** *that is, if c is any member of $\mathscr{R}$, then:*

$$(3) \qquad \sum_{k=m}^{n} cf(k) = c \sum_{k=m}^{n} f(k).$$

(iv) *If $f(n) = 1$ for every n:*

$$(4) \qquad \sum_{k=m}^{n} 1 = n - m + 1.$$

Proof. (i): For a given function f, the result of the addition in (1) depends only on the values of m and n. (ii): This is true by the general associative and commutative laws (§ 305), since the two members of (2) are simply two groupings of the same total set of terms. (iii): This follows from the general distributive law (§ 305). (iv): The left-hand member is the number of terms of the sum (1). If $m = 0$ this is $n + 1$. If $m = 1$, this is n. If $m \geq 2$, the number of terms in (1) is the number of natural numbers in $1, 2, \cdots, n$ minus the number of natural numbers in $1, 2, \cdots, m - 1$; in other words, $n - (m - 1) = n - m + 1$. (Cf. Theorem I, § 314, for the sum of an arithmetic sequence with common difference equal to 0.)

A useful summation formula is given in the theorem:

Theorem II. *If n is any natural number,*

(5)
$$\sum_{k=1}^{n} [f(k) - f(k - 1)] = f(n) - f(0).$$

Proof. Let $P(n)$ be the statement of the theorem. Then $P(1)$ is a triviality. Assuming that $P(n)$ is true for a given natural number, we wish to show that $P(n + 1)$ is also true. We write out the left-hand member of (5) for the case $n + 1$, as follows:

(6)
$$\sum_{k=1}^{n+1} [f(k) - f(k - 1)] = \sum_{k=1}^{n} [f(k) - f(k - 1)] + [f(n + 1) - f(n)].$$

By the induction assumption that $P(n)$ is true, the sum on the right of (6) can be replaced by the right-hand member of (5), so that (6) reduces to $[f(n) - f(0)] + [f(n + 1) - f(n)] = f(n + 1) - f(0)$, as desired. By the Fundamental Theorem of Mathematical Induction the proof is complete.

The following examples illustrate some of the uses of the preceding theorems. The word *derive* is used in the sense of *discover*, that is, *obtain without advance knowledge*.

Example 1. Derive the formula

(7)
$$1 + 2 + \cdots + n = \tfrac{1}{2}n(n + 1).$$

Solution. If $f(n) \equiv n^2$, formula (5) becomes

$$\sum_{k=1}^{n} [k^2 - (k - 1)^2] = \sum_{k=1}^{n} (2k - 1) = n^2.$$

By Theorem I, this can be rewritten as

$$\sum_{k=1}^{n} (2k) + \sum_{k=1}^{n} (-1) = 2 \sum_{k=1}^{n} k + (-1) \sum_{k=1}^{n} 1 = n^2,$$

or $2(1 + 2 + \cdots + n) - n = n^2$, and (7) follows. Since the left-hand member of (7) is the sum of n terms of an arithmetic sequence, the result can also be obtained from formula (10), § 314.

Example 2. Derive the formula

(8)
$$1^2 + 2^2 + \cdots + n^2 = \tfrac{1}{6}n(n + 1)(2n + 1).$$

Solution.   Let $f(n) \equiv n^3$.   Then, by formula (5),

$$\sum_{k=1}^{n} [k^3 - (k - 1)^3] = \sum_{k=1}^{n} [3k^2 - 3k + 1] = n^3,$$

and hence, by Theorem I, and equation (7):

$$3 \sum_{k=1}^{n} k^2 - 3(\tfrac{1}{2}n(n + 1)) + n = n^3,$$

and the left-hand member of (8) is equal to

$$\tfrac{1}{6}[2n^3 - 2n + 3n(n + 1)] = \tfrac{1}{6}n(n + 1)(2n + 1).$$

Example 3.   Derive the formula of Example 1, § 307:

(9) $1^2 + 3^2 + \cdots + (2n - 1)^2 = \tfrac{1}{3}n(4n^2 - 1).$

Solution.   The left-hand member of (9) can be written

$$\sum_{k=1}^{n} (2k - 1)^2 = \sum_{k=1}^{n} (4k^2 - 4k + 1) = 4 \sum_{k=1}^{n} k^2 - 4 \sum_{k=1}^{n} k + \sum_{k=1}^{k} 1,$$

and therefore, by (4), (7), and (8), is equal to

$$\tfrac{2}{3}n(n + 1)(2n + 1) - 2n(n + 1) + n = \tfrac{1}{3}n[4n^2 + 6n + 2 - 6n - 6 + 3].$$

This is the right-hand member of (9).

318. EXERCISES

In Exercises 1–2, find the sum of the given finite arithmetic sequence.

1. $18 + 30 + 42 + \cdots + 1818.$

2. $6 + 10 + 14 + \cdots + (4n + 2).$

In Exercises 3–4, find the sum of the given finite geometric sequence.

3. $2187 + 729 + 243 + \cdots + 1 + 1/3.$

4. $1 + 2 + 4 + \cdots + 2^{n-1}.$

5. Establish a one-to-one correspondence between the natural numbers and the even natural numbers, that is, those of the form $2n$ where n is a natural number, and prove that your correspondence is one-to-one.

6. Establish a one-to-one correspondence between the natural numbers and those natural numbers that belong to the range of the factorial function (cf. Example 4, § 310), and prove that your correspondence is one-to-one.

In Exercises 7–12, use the methods of § 317 to derive the formula of the specified exercise of § 308.

7. Ex. 13.   **8.** Ex. 14.   **9.** Ex. 15.

10. Ex. 16.   **11.** Ex. 17.   **12.** Ex. 18.

In Exercises 13–18, use the methods of § 317 to derive a formula for the given sum.

13. $2^2 + 5^2 + 8^2 + \cdots + (3n - 1)^2.$

14. $1^3 + 3^3 + 5^3 + \cdots + (2n - 1)^3$.

15. $1 \cdot 2 + 2 \cdot 3 + 3 \cdot 4 + \cdots + n(n + 1)$. (Cf. Ex. 28.)

16. $1^2 \cdot 2 + 2^2 \cdot 3 + 3^2 \cdot 4 + \cdots + n^2(n + 1)$.

17. $1 \cdot 2^2 + 2 \cdot 3^2 + 3 \cdot 4^2 + \cdots + n(n + 1)^2$.

18. $1^2 \cdot 2^2 + 2^2 \cdot 3^2 + 3^2 \cdot 4^2 + \cdots + n^2(n + 1)^2$.

19. Let $\{m_n\} = m_1, m_2, \cdots, m_n, \cdots$ be a strictly increasing sequence of natural numbers, that is, $m_{n+1} > m_n$ for every natural number n. Prove that $m_n \geq n$ for every natural number n.

20. Use mathematical induction to prove that $n^n > n!$ for every natural number n. (Cf. Example 4, § 310.)

21. Use mathematical induction to prove that $n! > 2^n$ for $n > 3$. (Cf. Exs. 22, 23.)

22. Use mathematical induction to prove that if m and N are natural numbers such that $N \geq m$ and $N! > m^N$ then the inequality $n \geq N$ implies $n! > m^n$. (Cf. Exs. 21, 23.)

23. If m is a natural number, prove that there exists a natural number N such that $N \geq m$ and $N! > m^N$. *Hint:* Let $N \equiv (2m)^{2m}$. Then (Example 2, § 307) $2^N > N$ and $N! = (1 \cdot 2 \cdots \cdots 2m)[(2m + 1)(2m + 2) \cdots \cdots N] > (2m)^{N-2m} > m^N$. (Cf. Exs. 21, 22.)

24. If A and B are finite sets, prove that the **union** $A \cup B$ of the two sets, defined to be the set of all points p such that $p \in A$ or $p \in B$ (or both), is finite. *Hint:* Let C be the set of all points of B that are not points of A. Then $A \cup C = A \cup B$, and A and C have no point in common. If A has m points and C has n points consider the two sets of natural numbers: $\{1, 2, \cdots, m\}$ and $\{m + 1, m + 2, \cdots, m + n\}$.

25. If n and r are any two numbers each of which is either 0 or a natural number and if $r \leq n$, then the corresponding **binomial coefficient** (cf. Exs. 26, 27) is defined:

(1)
$$\binom{n}{r} \equiv \frac{n!}{(n - r)! \, r!},$$

where the factorial symbol is defined in Example 4, § 310. Prove that if $1 \leq r \leq n$:

(2)
$$\binom{n + 1}{r} = \binom{n}{r - 1} + \binom{n}{r}$$

This is the famous law of **Pascal's triangle**, after B. Pascal (1623–1662, French), (cf. any good College Algebra text), which states that if the binomial coefficients are arranged in a triangular pattern, with $\binom{0}{0} = 1$ at the top vertex, with $\binom{1}{0} = 1$ and $\binom{1}{1} = 1$ on the next row, and so forth, thus:

$$
\begin{array}{c}
1 \\
1 \quad 1 \\
1 \quad 2 \quad 1 \\
1 \quad 3 \quad 3 \quad 1 \\
1 \quad 4 \quad 6 \quad 4 \quad 1
\end{array}
$$

$\cdots$

then every entry of every row, except for the extreme 1's, is equal to the sum of the two entries nearest to it in the preceding row (the row immediately above).

26. Prove that every binomial coefficient (1) is a natural number. (Cf. Exs. 25, 27.)

27. Prove the **Binomial Theorem**, where x and y are arbitrary real numbers (or, more generally, members of an arbitrary field) and n is a natural number:

$$(3) \quad (x + y)^n = \binom{n}{0}x^n + \binom{n}{1}x^{n-1}y + \cdots + \binom{n}{r}x^{n-r}y^r + \cdots + \binom{n}{n}y^n.$$

Hint: Use mathematical induction, multiplying (3) by $(x + y)$ and using (2).

28. Show that the formula of Exercise 15 can be expressed:

$$\binom{2}{2} + \binom{3}{2} + \cdots + \binom{n + 1}{2} = \binom{n + 2}{3}.$$

(Cf. Exs. 25–27.)

29. If m is a fixed but arbitrary natural number, show that the sum

$$\sum_{k=1}^{n} k^m = 1^m + 2^m + 3^m + \cdots + n^m$$

can be written in the form

$$\frac{n^{m+1}}{m + 1} + a_m n^m + a_{m-1}n^{m-1} + \cdots + a_2 n^2 + a_1 n + a_0,$$

for every natural number n. (Cf. Exs. 11–15, § 308.) *Hint:* Use the method of § 317, with induction on m, with the aid of the binomial theorem (Ex. 27).

30. Prove the **pigeon-hole principle**: Let $x_1, x_2, \cdots, x_m$ be any m points, let $A_1, A_2, \cdots, A_n$ be any n sets, and assume that for each $j = 1, 2, \cdots, m$ the statement $x_j \in A_k$ is true for at least one $k = 1, 2, \cdots, n$; then, if $n < m$ there must exist two x's with distinct subscripts that belong to the same A_k; that is, there must exist natural numbers i, j, and k such that $1 \leq i \leq m$, $1 \leq j \leq m$, $1 \leq k \leq n$, $i \neq j$, $x_i \in A_k$, and $x_j \in A_k$. This is named as it is since, in the special case that the x's are m distinct objects and no two of the A's have a point in common, the A's can be thought of as representing containers or "pigeon holes" for these objects. If the objects are placed individually into the pigeon-holes and there are too few pigeon-holes to "go around," then at least two of the objects must find themselves in the same pigeon-hole.

31. If $\{a_n\}$ is the Fibonacci sequence of the Example, § 315, prove that $a_n < 2^{n-1}$ for every $n > 1$.

32. Let $\{a_n\}$ and $\{b_n\}$ be two sequences of natural numbers defined inductively: $a_1 = 1$, $b_1 = 1$; for $n > 1$, $a_n = a_{n-1} + b_{n-1}$, $b_n = a_n b_{n-1}$. Prove that these sequences are well defined, and that $a_n \leq 2^{2^{n-2}}$ for $n > 1$, and $b_n \leq 2^{2^{n-1}-1}$ for all n.

4

Composite and Prime Numbers

++

401. DIVISORS AND MULTIPLES

Definition I. *A natural number m **divides** a natural number k (in symbols: m | k) if and only if there exists a natural number n such that k = mn. If m | k, m is called a **divisor** or **factor** of k, and k is called a **multiple** of m. A natural number k is called **composite** if and only if there exist natural numbers m and n such that m > 1, n > 1, and k = mn. A natural number p is **prime** (or **a prime**) if and only if p > 1 and p is not composite.*

Examples. Every natural number n is a divisor of itself, $n \mid n$, since there exists the natural number 1 such that $n = n \cdot 1$. Similarly, 1 is a divisor of every natural number n, $1 \mid n$, since $n = 1 \cdot n$. The following numbers are composite: 4, 6, 100, and 1001. The following numbers are prime: 2, 3, 5, 7, 11, 13, and 101.

Theorem I. *Let k, m, and n denote natural numbers. If $k \mid m$ and if $m \mid n$ then $k \mid n$. If $k \mid m$ then $k \mid mn$. If $k \mid m$ and if $k \mid n$ then $k \mid (m + n)$. If $k \mid m$, if $k \mid n$, and if $m < n$ then $k \mid (n - m)$. If $k \mid m$ then $k \leq m$. If $k \mid m$ and $m \mid k$, then $k = m$. If $k \mid m$ and if k and m are prime, then $k = m$.*

Proof. If $k \mid m$ and $m \mid n$, there exist natural numbers a and b such that $m = ak$ and $n = bm = b(ak) = (ba)k$, so that $k \mid n$. If $k \mid m$ there exists a natural number a such that $m = ak$, so that $mn = (ak)n = (an)k$, so that $k \mid mn$. If $k \mid m$ and $k \mid n$ there exist natural numbers a and b such that $m = ak$ and $n = bk$, so that $m + n = (a + b)k$, and $k \mid (m + n)$. If, in addition, $m < n$, then $n - m = bk - ak = (b - a)k$ is positive, so that $b - a > 0$ (cf. § 203), and hence $b > a$. Therefore $b - a$ is a natural number (V, § 304), and $k \mid (n - m)$. If $k \mid m$ there exists a natural number a such that $m = ak$. Since $a \geq 1$, $m = ak \geq 1 \cdot k = k$ (cf. Ex. 6, § 204). If $k \mid m$ and $m \mid k$, then $k \leq m \leq k$, and hence $k = m$. Finally, if $k \mid m$ and if k and m are prime, $1 < k \leq m$, and $m = nk$ for some natural number n; but if n were greater than 1, m would be composite; therefore $n = 1$ and $k = m$.

The next two theorems establish the existence of prime divisors.

Theorem II. *Every natural number greater than 1 has a least (or smallest) divisor greater than 1. This least divisor is prime.*

Proof. Let n be a natural number greater than 1, and let A be the set of all natural numbers greater than 1 that are divisors of n. Since n is a member of A, A is not empty. By the well-ordering principle (VII, § 304), A has a least member p. To show that p is prime, assume that it is composite: $p = ab$, where a and b are natural numbers greater than 1. Then, by Theorem I, since $a \mid p$, $a \mid n$. Furthermore, also by Theorem I, $a \leq p$, and if a were equal to p, b would be equal to 1, contrary to assumption. Therefore $1 < a < p$, and a is a member of the set A less than its least member. With this contradiction the proof is complete.

An immediate corollary is the theorem:

Theorem III. *Every natural number greater than 1 has at least one prime divisor.*

Theorem III puts us in position to reproduce a proof of the infinitude of the primes, attributed to Euclid (365?–275? B.C.):

Theorem IV. *There are infinitely many prime numbers.*

Proof. Assume there are only finitely many distinct primes: $p_1, p_2, \cdots, p_n$. Let q designate their product: $q = p_1 p_2 \cdots p_n$, and let $r = q + 1$. By the Corollary to Theorem II, the natural number r has a prime divisor p, which by assumption must be one of the finite set $p_1, p_2, \cdots, p_n : p = p_k$, where $1 \leq k \leq n$. We now have, simultaneously, $p \mid r$ and $p \mid q$ (since $p = p_k$ is one of the factors of q), and therefore, by Theorem I, $p \mid (r - q)$. But this means that $p \mid 1$, which is impossible by Theorem I since $p > 1$.

We conclude this section with a fundamental fact about the factorability of natural numbers into products of primes:

Theorem V. *Every natural number greater than 1 is a product of a finite number of primes (finite repetitions being permitted). That is, if n is a natural number such that $n > 1$, then there exist prime numbers $p_1, p_2, \cdots, p_m$, where $m \geq 1$, such that $n = p_1 p_2 \cdots p_m$.*

Proof. Let A be the set of all divisors a of the natural number n ($n > 1$) that have the form of a product of primes: $a = p_1 p_2 \cdots p_m$ ($m \geq 1$). By Theorem III, A is nonempty. Since $1 \leq a \leq n$ for every member a of A, A is finite and has a greatest member k (Example 4, § 306). We wish to show that $k = n$. Assume $k < n$. Then, since $k \mid n$, there exists a natural number q greater than 1 such that $n = kq$. Let p be a prime divisor of q, so that $q = pr$. Then $n = k(pr) = (kp)r$, and kp is a product of primes and a divisor of n. That is, kp is a member of A. But since $p > 1$, $kp > k$, where k is the *greatest* member of A. With this contradiction the proof is complete.

402. LOWEST TERMS

Definition. *Two natural numbers are **relatively prime** if and only if the only natural number that divides both of them is unity; that is, if and only if they have no common divisor greater than 1. A fraction a/b, where the numerator a and the denominator b are natural numbers, is in **lowest terms**, if and only if a and b are relatively prime.*

A convenient criterion for relative primeness is the following:

Theorem I. *Two natural numbers are relatively prime if and only if they have no common prime divisor.*

Proof. If two numbers a and b have a common divisor d greater than 1, let p be a prime divisor of d (Theorem III, § 401); then p is a common prime divisor of a and b. On the other hand, any common prime divisor of a and b is a common divisor greater than 1.

An important fact concerning fractions is that they can be reduced to lowest terms.

Theorem II. *Any given fraction whose numerator and denominator are natural numbers is equal to a fraction in lowest terms. That is, if a and b are natural numbers there exist relatively prime natural numbers c and d such that $a/b = c/d$.*

Proof. Let $P(n)$ be the proposition that the theorem is true whenever $a \leq n$ and $b \leq n$. Then $P(1)$ is trivial. Assuming the truth of $P(n)$, for a specific natural number n, we shall seek a fraction c/d in lowest terms equal to a given fraction a/b, where $a \leq n + 1$ and $b \leq n + 1$. If a and b are relatively prime, let $c = a$ and $d = b$. Otherwise, by Theorem I, there exists a common prime divisor p of a and b. If $a = rp$ and $b = qp$, we can write the fraction $a/b = rp/qp$ in the reduced form r/q. Since $rp = a \leq n + 1 \leq n + n = 2n \leq pn$ it follows that $r \leq n$. Similarly, $q \leq n$. Therefore, by the induction assumption that $P(n)$ is true, the fraction r/q is equal to a fraction c/d in lowest terms. Finally, since $a/b = r/q = c/d$, an application of the Fundamental Theorem of Mathematical Induction completes the proof.

NOTE. We have not yet proved that the reduction of a fraction to lowest terms produces a unique result. The uniqueness of a fraction in lowest terms requires more substantial machinery than we have at our disposal presently, and is established in Theorem V, § 404. In § 1009 an example of a number system is given where there exist fractions in lowest terms that are not unique.

403. THE FUNDAMENTAL THEOREM OF EUCLID

Our first step in this section is to show that the set of natural numbers has what is known as the *Archimedean property*. It will be shown later (§ 903) that the real number system is also Archimedean.   (Also, cf. the Note, § 501, Ex. 8, § 506, and § 713.)

Theorem I.   Archimedean Property.   *If a and b are natural numbers there exists a natural number n such that $na > b$.*

Proof.   Let $n \equiv b + 1$.   Then $na = a(b + 1) = ab + a > ab \geqq b$.

Theorem II.   Fundamental Theorem of Euclid.   *If a and b are natural numbers there exist unique numbers n and r, each of which is either 0 or a natural number, such that*

(1) $$r < a$$

and

(2) $$b = na + r.$$

Proof.   For existence, let m be the least natural number such that $ma > b$ (this exists, by Theorem I and the Well-Ordering Principle, VII, § 304). If $n \equiv m - 1$, then n is either 0 or a natural number.   If $n = 0$, then $m = 1$ and $a > b$, so that if r is taken to be equal to b, r will satisfy both (1) and (2). If $n \neq 0$, then n is a natural number so that (by the definition of m as the *least* natural number such that $ma > b$) $na \leqq b$.   If $na = b$, let r be 0.   If $na < b$, let r be the natural number $r \equiv b - na$.   Then r satisfies equation (2).   The inequality (1) follows from the fact that $ma > b$: $(n + 1)a = na + a > b$, and hence $r = b - na < a$.

For uniqueness, assume $b = n_1 a + r_1 = n_2 a + r_2$, where n_1, n_2, r_1, and r_2 are numbers each of which is either 0 or a natural number, and where $r_1 < a$ and $r_2 < a$.   If $r_1 = r_2$, then the equality $n_1 a + r_1 = n_2 a + r_2$ implies the equality $n_1 a = n_2 a$, and hence $n_1 = n_2$, and uniqueness is proved.   Assume, therefore, that r_1 and r_2 are not equal, and adjust the notation so that $r_1 < r_2$. Then, by subtraction, we infer, from the equality $n_1 a + r_1 = n_2 a + r_2$, that

(3) $$n_1 a - n_2 a = r_2 - r_1.$$

The inequalities $0 \leqq r_1 < r_2 < a$ imply that $0 < r_2 - r_1 \leqq r_2 < a$, and hence that $0 < (n_1 - n_2)a < a$.   But this means that $n_1 - n_2$ is a natural number less than 1, and the desired contradiction has been obtained.

NOTE 1.   The number r of Theorem II is called the **remainder** when b is divided by a.   The number n is called the **quotient,** and b and a are called the **dividend** and the **divisor**, respectively.   The word *quotient*, therefore, has two distinct and contradictory definitions, that of § 102 and the present one.   The context should in all cases dispel any possible misinterpretation.   For definiteness, in this book (unless explicit statement to the contrary is made) *the single word quotient should be interpreted to mean a ratio, as defined in § 102.*

NOTE 2.   If $a = 2$, the Fundamental Theorem of Euclid states that any natural number b has either the form $b = 2n$ or the form $b = 2n + 1$, where n is either 0 or a natural number.   If b has the form $2n$ it is called **even**, and if b has the form $2n + 1$ it is called **odd**.

404. DIVISIBILITY OF PRODUCTS

The Fundamental Theorem of Euclid (§ 403) provides a useful piece of machinery for obtaining a number of interesting results concerning the divisibility of a product of natural numbers by a prime. Basic to these is the theorem:

Theorem I. *A prime number p cannot be a divisor of the product of two natural numbers a and b each of which is less than p.*

Proof. Assume that p is a prime divisor of ab, where a and b are natural numbers such that $a < p$ and $b < p$. Let c be the *least* natural number such that $p \mid ac$. Then $1 < c \leq b < p$. Let n and r be natural numbers, according to the Fundamental Theorem of Euclid (§ 403), such that $p = nc + r$, where $0 < r < c$ (neither n nor r can vanish since $1 < c < p$ and p is prime). Then $ar = ap - n(ac)$, which is divisible by p. In other words, r is a natural number less than c such that $p \mid ar$, whereas c was assumed to be the least such natural number! With this contradiction the proof is complete.

Theorem II. *If p is a prime number, if a and b are natural numbers, and if $p \mid ab$, then $p \mid a$ or $p \mid b$ (or both).*

Proof. Let $a = mp + r$ and $b = np + s$, where m, n, r, and s are numbers each of which is either 0 or a natural number, and $r < p$ and $s < p$. Assume that $p \mid ab$ and that neither r nor s is equal to 0. Then, since $ab = (mnp + ms + nr)p + rs$ it follows that $p \mid rs$. But this contradicts Theorem I. Therefore, if $p \mid ab$ it must follow that either $r = 0$ or $s = 0$. If $r = 0$ then $p \mid a$, and if $s = 0$ then $p \mid b$, and the proof is complete.

The following theorem extends the preceding result to an arbitrary number of factors:

Theorem III. *A prime number p that divides the product of n natural numbers must divide at least one of the n factors.*

Proof. Let $P(n)$ be the statement of the theorem for the natural number n. $P(1)$ is trivial and $P(2)$ is Theorem II. Assuming $P(n)$, for a fixed n, we wish to prove $P(n + 1)$. Assume, therefore, that $p \mid a_1a_2 \cdots a_na_{n+1}$. Then either $p \mid a_1a_2 \cdots a_n$ or $p \mid a_{n+1}$, by Theorem II, since the product of the $n + 1$ factors can be written as the product of two factors: $(a_1a_2 \cdots a_n)$ and (a_{n+1}). If $p \mid a_1a_2 \cdots a_n$, then by the induction assumption p divides at least one a_k, $1 \leq k \leq n$. On the other hand, if $p \mid a_{n+1}$ there is no more to prove.

We now extend Theorem II in another direction, by removing the restriction that the divisor p be prime:

Theorem IV. *If a, b, and c are natural numbers, if c and a are relatively prime, and if $c \mid ab$, then $c \mid b$.*

Proof. Since the statement is trivial for $c = 1$ we assume henceforth that $c > 1$, and obtain a proof by induction on the number of prime factors in a possible factorization of c. That is, let $P(n)$ be the proposition that the theorem is true whenever c can be represented as a product of n primes (cf. Theorem V, § 401). In the first place, $P(1)$ is true by Theorem II, since in this case c is a prime and c does not divide a. Assume that $P(n)$ is true for a certain natural number n, and let c be a product of $n + 1$ prime factors: $c = p_1 p_2 \cdots p_{n+1}$. Assuming that a and c have no common prime divisors and that $c \mid ab$ (that is, $ab = cm$ for some natural number m), we infer that p_{n+1} is a divisor of ab and not of a, and hence $p_{n+1} \mid b$. Let $c' = p_1 p_2 \cdots p_n$, so that $c = c' p_{n+1}$, and let $b = b' p_{n+1}$. Then c' and a are relatively prime, $c' \mid ab'$ (since $ab' = c'm$), and c' is a product of n primes. Hence, by the induction hypothesis, $c' \mid b'$. But this means that there exists a natural number d such that $b' = c'd$. Multiplying both members by p_{n+1} gives $b = cd$, and hence $c \mid b$, as desired. Since every natural number greater than 1 can be represented as a product of n primes for some natural number n (Theorem V, § 401), an application of the Fundamental Theorem of Mathematical Induction completes the proof.

Theorem IV enables us to prove that the result of reducing a fraction to lowest terms is unique.

Theorem V. *If a/b and c/d, where a, b, c, and d are natural numbers, are fractions in lowest terms, and if $a/b = c/d$, then $a = c$ and $b = d$.*

Proof. From the equation $a/b = c/d$ we obtain the equation $ad = bc$, and hence $a \mid bc$ and $c \mid ad$. From Theorem IV, since a and b are relatively prime we infer that $a \mid c$, and since c and d are relatively prime we infer $c \mid a$. Therefore $a = c$, and hence, from $ad = bc$, $b = d$.

405. UNIQUE FACTORIZATION THEOREM

The theorem of this section, which was given in Euclid's *Elements*, is so basic to all of arithmetic and number theory that it is often called the *Fundamental Theorem of Arithmetic*. It is a uniqueness theorem for the representation of a natural number greater than 1 as a product of primes, guaranteed to exist by Theorem V, § 401. (Also cf. § 707.) For an example of a number system where unique factorization *fails*, see § 1009. For further discussion of this topic, see G. Birkhoff and S. MacLane, *A Survey of Modern Algebra* (New York, The Macmillan Company, 1953), pp. 21, 75, 416, 425.

Theorem. Unique Factorization Theorem. *The representation of a natural number greater than 1 as a product of primes is unique except for the order of the factors. If the prime factors are arranged according to their order relationship (cf. Example 8, § 310) the representation is unique. In other words, if $p_1, p_2, \cdots, p_r$ and $q_1, q_2, \cdots, q_s$ are prime numbers, if $p_1 \leqq p_2 \leqq \cdots \leqq p_r$ and $q_1 \leqq q_2 \leqq \cdots \leqq q_s$, and if*

(1) $$p_1 p_2 \cdots p_r = q_1 q_2 \cdots q_s,$$

then $r = s$ and $p_k = q_k$ for $1 \leqq k \leqq r = s$.

Proof. Let $P(n)$ be the statement of the last sentence of the Theorem, with the added conditions that $r \leq n$ and $s \leq n$. Then $P(1)$ is a triviality. For $n = 2$, equation (1) can take only the following four forms: $p_1 = q_1, p_1 p_2 = q_1$, $p_1 = q_1 q_2$, and $p_1 p_2 = q_1 q_2$. The first of these is one of the two alternatives sought, and the next two are impossible since no prime is composite. This brings us to the equation $p_1 p_2 = q_1 q_2$, which we now assume to be true. By Theorem II, § 404, since $p_2 \mid q_1 q_2$, either $p_2 \mid q_1$ or $p_2 \mid q_2$, or both. Therefore either $p_2 \leq q_1$ or $p_2 \leq q_2$, or both, and since $q_1 \leq q_2$, it follows that in any case $p_2 \leq q_2$. By similar reasoning we find $q_2 \leq p_2$, and hence $p_2 = q_2$. Consequently $p_1 = q_1$, and $P(2)$ is proved. We now assume $P(n)$ to be true for a certain natural number n, and seek the establishment of $P(n + 1)$. Accordingly, assume an equation of the form (1) to be true, where $r \leq n + 1$ and $s \leq n + 1$. Since p_r is a divisor of the right-hand member of (1), $p_r \mid q_k$ for some k where $1 \leq k \leq s$, by Theorem III, § 404. Since $p_r \leq q_k$ and $q_k \leq q_s$ we have $p_r \leq q_s$. By a similar process, $q_s \leq p_r$ and we infer that $p_r = q_s$. Dividing both members of (1) by $p_r = q_s$, we have the equality $p_1 p_2 \cdots p_{r-1} = q_1 q_2 \cdots q_{s-1}$. By the induction assumption, since $r - 1 \leq n$ and $s - 1 \leq n$, we conclude that $r - 1 = s - 1$ (and hence $r = s$), and that $p_k = q_k$ for $1 \leq k \leq r - 1$ (and hence for $1 \leq k \leq r$). With this, and an application of the Fundamental Theorem of Mathematical Induction, the proof is complete.

406. EXERCISES

1. Prove that there is exactly one even prime number.

2. Let n be a natural number. Prove that n^2 is even if and only if n is even. Prove that n^2 is odd if and only if n is odd.

3. Prove that a natural number n is prime if and only if its only prime factor is n.

4. Prove that in determining whether a natural number n is prime the only divisibility tests that need be made are those of the divisibility of n by the primes 2, 3, 5, 7, $\cdots$, p, where $p^2 \leq n$ and $(p + 1)^2 > n$. (For example, if $n = 1003, p = 31$.)

In Exercises 5–8, find the factorization into prime factors of the given number.

5. 1927.   **6.** 1999.

7. 1,548,547.   **8.** 871,933.

9. Define $p_1 = 2, p_2 = 3, p_3 = 5$, etc., where, for every natural number n, p_{n+1} is defined to be the least prime greater than p_n. Prove that the sequence $\{p_n\}$ of primes is well defined. Write out the first 30 primes.

10. Actually carry out Euclid's method of finding a prime greater than p_n, where p_n is the nth prime (cf. Ex. 9), by considering the prime factors of $(p_1 p_2 \cdots p_n) + 1$, for the cases $n = 1, 2, 3, 4$, and 5.

11. By considering the sequence $n! + 2, n! + 3, \cdots, n! + n$, show that there exist arbitrarily long sequences of consecutive natural numbers all of which are composite.

12. Show that there exists one and only one pair of prime numbers p and q such that $p - q = 11$.

13. Prove that whenever a natural number n is represented as a product of m natural numbers each of which is greater than 1, then $n > m$.

14. If m and n are natural numbers and if $11 \mid (5m + 4n)$, prove that $11 \mid (m + 3n)$.

15. Two prime numbers that differ by 2 are called a pair of **twin primes.** (It has been conjectured that there are infinitely many pairs of twin primes, but this has never been proved.)   Find the first pair of twin primes greater than 200.

16. In 1742, C. Goldbach (1690–1764, Russian) conjectured that every even number greater than 2 is the sum of two primes.   (The Goldbach conjecture has never been proved.)   Write each of 20, 30, 40, 50, and 60 as the sum of two primes in all possible ways.

17. Show that for $n = 1, 2, 3, 4, \cdots, 40$, $n^2 - n + 41$ is prime.   Is $n^2 - n + 41$ prime for every natural number n?   Can you find infinitely many n such that $n^2 - n + 41$ is composite?   Can you find infinitely many n such that n and 41 are relatively prime and $n^2 - n + 41$ is composite?

18. If $\{a_n\}$ is the Fibonacci sequence of the Example, § 315, prove that for every natural number n, a_n and a_{n+1} are relatively prime.

19. If a, b, and c are natural numbers, if a and c are relatively prime, and if b and c are relatively prime, prove that ab and c are relatively prime.   If a and c are relatively prime and n is a natural number, prove that a^n and c are relatively prime.

20. If a, b, and c are natural numbers, if a and b are relatively prime, if $a \mid c$, and if $b \mid c$, prove that $ab \mid c$.

*407. THE EUCLIDEAN ALGORITHM

Any system of calculation that consists of a routine repetition of a simple basic step is called an **algorithm,** or **algorism.**   An example is the standard method of long division (cf. §§ 1202 and 1208).   Another is the square root algorithm (discussed in § 1210).

One of the most useful algorithms of arithmetic—called the *Euclidean Algorithm*—consists essentially of repeated application of the Fundamental Theorem of Euclid.   Let a and b be any two natural numbers and assume for definiteness that $a \geq b$.   By the Fundamental Theorem of Euclid (§ 403), there exist a natural number q_1 and a number r_1 that is either 0 or a natural number such that $r_1 < b$ and $a = q_1 b + r_1$.   If $r_1 > 0$ the division process can be repeated, this time with b and r_1: $b = q_2 r_1 + r_2$, where q_2 is a natural number and r_2 is a number that is either 0 or a natural number, and where $r_2 < r_1$. If $r_2 > 0$, repeat the process with r_1 and r_2: $r_1 = q_3 r_2 + r_3$, etc.   Since $b > r_1 > r_2 > r_3 > \cdots$, this process must terminate (cf. the Theorem, § 316).   That is, there must be a natural number $k + 1$ such that $r_{k-1} = q_{k+1} r_k + 0$.   We now display the Euclidean Algorithm in summary:

---

* Starred sections and chapters may be omitted without destroying the continuity of the remaining material.   See the preface for a discussion of the role of starring in this book.

Theorem. Euclidean Algorithm. *If a and b are natural numbers, with* $a \geq b$, *and if b does not divide a, then there exist natural numbers* $q_1, q_2, \cdots ,$ q_{k+1} *and* $r_1, r_2, \cdots , r_k$, *where* $k \geq 1$, *such that*

(1)
$$
\begin{cases}
a = q_1 b + r_1, & 0 < r_1 < b, \\
b = q_2 r_1 + r_2, & 0 < r_2 < r_1, \\
r_1 = q_3 r_2 + r_3, & 0 < r_3 < r_2, \\
\qquad \cdots \\
r_{k-2} = q_k r_{k-1} + r_k, & 0 < r_k < r_{k-1}, \\
r_{k-1} = q_{k+1} r_k.
\end{cases}
$$

In case $k = 1$, *the last equation of* (1) *should be interpreted as* $r_0 \equiv b = q_2 r_1$. *In case* $b \mid a$, *system* (1) *reduces to the single equation* $a = q_1 b$ *which, with* $r_{-1} \equiv a$ *and* $r_0 \equiv b$, *is the last equation of* (1) *with* $k = 0$.

Example. Apply the Euclidean Algorithm to the numbers 9035 and 364.

Solution. Successive divisions give:

(2)
$$
\begin{cases}
9035 = 24 \cdot 364 + 299, \\
364 = 1 \cdot 299 + 65, \\
299 = 4 \cdot 65 + 39, \\
65 = 1 \cdot 39 + 26, \\
39 = 1 \cdot 26 + 13, \\
26 = 2 \cdot 13.
\end{cases}
$$

*408. GCD AND LCM

If a and b are any two given natural numbers, the set C of all natural numbers c that are common divisors of both a and b (that is, $c \mid a$ and $c \mid b$) is a nonempty finite set ($1 \in C$ and $1 \leq c \leq a$). The set C therefore has a greatest member d (Example 4, § 306), called the *greatest common divisor* of a and b. This number d has a more remarkable property than merely being the greatest member of C. This property is incorporated into the following definition, and proved to exist in Theorem I below.

Definition I. *A natural number d is called the* **greatest common divisor**, *or* GCD *for short, of the natural numbers a and b if and only if it has the following two properties:*
  (i) $d \mid a$ *and* $d \mid b$;
  (ii) whenever c is a natural number and $c \mid a$ *and* $c \mid b$, *then* $c \mid d$.
The GCD *of a and b is denoted†*

(1)   *The* GCD *of a and b* $= d = (a, b)$.

† The notation (a, b) for the GCD of a and b is the same as the notation for the ordered pair (a, b) introduced in § 309. The context should make it amply clear which interpretation should be attached to the symbol (a, b), and no confusion should result from this double use of parentheses. (The GCD of two natural numbers is sometimes also denoted HCF, for *highest common factor*.)

Theorem I. *If a and b are any two natural numbers, their GCD exists and is unique. In the notation of the Euclidean Algorithm* (Theorem, § 407), $(a, b) = r_k$.

Proof. Uniqueness is easy: if d_1 and d_2 are natural numbers having properties (*i*) and (*ii*), then $d_1 \mid d_2$ and $d_2 \mid d_1$, so that $d_1 = d_2$. To establish existence we assume for definiteness that $a \geq b$. The case $b \mid a$ is trivial, with $d = b$. If $a > b$, we shall now show that the number $d \equiv r_k$ of (1), § 407, has the two properties (*i*) and (*ii*). In the first place, by the last equation of (1), § 407, $d \mid r_k$ and $d \mid r_{k-1}$. Therefore, by the next-to-the-last equation of (1), § 407, $d \mid r_{k-2}$. By mathematical induction (proceeding upward through equations (1), § 407), d is a divisor of the left-hand member of each equation of (1), including both b and a, and (*i*) is proved. To prove (*ii*), we let c be an arbitrary common divisor of a and b. By the first equation of (1), § 407, written in the form $r_1 = a - q_1 b$, we see that $c \mid r_1$. Therefore, by the second equation of (1), § 407, written in the form $r_2 = b - q_2 r_1$, $c \mid r_2$. Again by induction, this time proceeding downward through equations (1), § 407, $c \mid r_j$ for every $j = 1, 2, \cdots, k$, and in particular $c \mid r_k$, or $c \mid d$, as desired. This completes the proof.

Example 1. By the Example, § 407,

$$(9035, 364) = 13.$$

The following two theorems are easily established, and their proofs are left to the reader.

Theorem II. *Two natural numbers a and b are relatively prime if and only if their GCD is 1:* $(a, b) = 1$.

Theorem III. *If a and b are natural numbers, if $d = (a, b)$, and if $a = a_1 d$ and $b = b_1 d$, then a_1 and b_1 are relatively prime:* $(a_1, b_1) = 1$.

Any two natural numbers a and b possess common multiples as well as common divisors—for example, their product ab is a multiple of each. As might be expected, among the common multiples of a and b is one of special significance, called the *least common multiple* (or sometimes the *lowest common multiple*), according to the definition:

Definition II. *A natural number e is called the **least common multiple**, or LCM for short, of the natural numbers a and b if and only if it has the following two properties:*
(*i*) $a \mid e$ *and* $b \mid e$;
(*ii*) *whenever c is a natural number and $a \mid c$ and $b \mid c$, then $e \mid c$.*
The LCM of a and b is denoted

(2)   *The LCM of a and b* $= e = [a, b]$.

Theorem IV. *If a and b are any two natural numbers, their* LCM *exists and is unique. The* GCD *and* LCM *of a and b are related:*

(3) $$(a, b)[a, b] = ab.$$

Proof. Let $d \equiv (a, b)$ and $e \equiv (ab)/d$. Write $a = a_1 d$ and $b = b_1 d$, where a_1 and b_1 are natural numbers. Then, since $e = ab_1 = a_1 b$, e is a natural number that is a multiple of both a and b. Now let c be an arbitrary common multiple of a and b, and write $c = ma = nb$. If this last equation is written in the form $ma_1 d = nb_1 d$ we obtain $ma_1 = nb_1$. Since a_1 and b_1 are relatively prime (Theorem III) and since $a_1 \mid nb_1$, we infer from Theorem IV, § 404, that $a_1 \mid n$, so that n can be written in the form $n = a_1 r$, where r is a natural number. Substitution in the equation $c = nb$ leads to:

$$c = a_1 rb = re.$$

Therefore $e \mid c$, as we wished to show.

Example 2. By Example 1 and Theorem IV,

$$[9035, 364] = (9035 \cdot 364)/13 = 9035 \cdot 28 = 252{,}980.$$

A useful by-product of the Euclidean Algorithm is a proof of the fact that the GCD of two natural numbers a and b can be expressed in a simple "linear" fashion in terms of a and b:

Theorem V. *If a and b are natural numbers and if d is their* GCD, *then there exist natural numbers m, n, r, and s such that:*

(3) $$ma - nb = rb - sa = d.$$

The natural numbers a and b are relatively prime if and only if there exist natural numbers m, n, r, and s such that

(4) $$ma - nb = rb - sa = 1.$$

Proof. We start by proving a lemma.

Lemma. *If a and b are natural numbers, then any natural number that can be expressed in the form ma − nb (where m and n are natural numbers) can also be expressed in the form rb − sa (where r and s are natural numbers), and conversely.*

Proof of Lemma. If a, b, m, and n are natural numbers, then $r = ma + na - n$ and $s = mb + nb - m$ are natural numbers, and $ma - nb = rb - sa$.

Proof of Theorem II, continued. By the symmetry of the statement of the theorem, we may assume without loss of generality that $a \geq b$. Our objective will be to show that there exist natural numbers m and n such that $(a, b) = ma - nb$; the existence of r and s will follow from the lemma. We look at the first few possible values of k, in the Euclidean Algorithm (1), § 407. If $k = 0$, then $b \mid a$, and $(a, b) = b = 2a - (2q_1 - 1)b$. If $k = 1$, then $(a, b) = r_1 = 1 \cdot a - q_1 b$. If $k = 2$, then $(a, b) = r_2 = b - q_2 r_1 = (1 + q_1 q_2)b - q_2 a$, and by the Lemma this has the form $ma - nb$. Now let $P(v)$ be the proposition that for each $u = -1, 0, 1, \cdots, v, r_u$ can be expressed in the form $ma - nb$. Then for any $v \geqq 3$, $r_v = r_{v-2} - q_v r_{v-1}$, and by the induction assumption and the Lemma, r_{v-2} can be written in the form $ma - nb$, and r_{v-1} can be written in the form $rb - sa$. Therefore $r_v = (ma - nb) - q_v(rb - sa) = (m + q_v s)a - (n + q_v r)b$, which is of the form sought. By the Fundamental Theorem of Mathematical Induction the formula (3) is obtained. The only part of the proof remaining to be established is that (4) is a sufficient condition that a and b be relatively prime. This is true since any common divisor of a and b, in the presence of equation (4), must also be a divisor of the right-hand member 1.

Example 3. Find natural numbers $m, n, r,$ and s such that

$$9035m - 364n = 364r - 9035s = 13.$$

Solution. By (2) of the Example, § 407, $13 = 39 - 1 \cdot 26 = 39 - (65 - 1 \cdot 39) = 2 \cdot 39 - 1 \cdot 65 = 2(299 - 4 \cdot 65) - 65 = 2 \cdot 299 - 9 \cdot 65 = 2 \cdot 299 - 9(364 - 1 \cdot 299) = 11 \cdot 299 - 9 \cdot 364 = 11(9035 - 24 \cdot 364) - 9 \cdot 364 = 9035 \cdot 11 - 364 \cdot 273$. This gives the values $m = 11$ and $n = 273$. To find r and s we write $9035 \cdot 11 - 364 \cdot 273 = 364(9035t - 273) - 9035(364t - 11)$ and let $t = 1$: $364 \cdot 8762 - 9035 \cdot 353$, with $r = 8762$ and $s = 353$. A simpler solution for r and s is given by assigning the value $1/13$ (the reciprocal of the GCD of 9035 and 364) to t, with $r = 422$ and $s = 17$.

Example 4. Find natural numbers $m, n, r,$ and s such that

$$77m - 36n = 36r - 77s = 1.$$

Solution. The Euclidean Algorithm gives:

$$\begin{cases} 77 = 2 \cdot 36 + 5, \\ 36 = 7 \cdot 5 + 1. \end{cases}$$

From this we see that 77 and 36 are relatively prime, and $1 = 36 - 7 \cdot 5 = 36 - 7(77 - 2 \cdot 36) = 36 \cdot 15 - 77 \cdot 7$. This gives the values $r = 15$ and $s = 7$. To find m and n we write $36 \cdot 15 - 77 \cdot 7 = 77(36t - 7) - 36(77t - 15)$, and let $t = 1$: $77 \cdot 29 - 36 \cdot 62$, with $m = 29$ and $n = 62$.

NOTE. If one examines the proof of the Lemma of Theorem V, and the solutions of Examples 3 and 4, it should be clear that there are infinitely many possibilities for the values of $m, n, r,$ and s in Theorem V. We have made no effort to find all solutions, or even to show that the solutions obtained in Examples 3 and 4 are the smallest possible. Such questions as these, relating to solutions involving natural numbers for equations in more than one unknown, are a part of one of the oldest and most fascinating chapters of number theory, called *Diophantine Equations*, after the Greek mathematician Diophantus (circa 250 A.D.). (Cf. Ex. 19, § 409.) For further reading, cf. B. W. Jones, *The Theory of Numbers* (New York, Rinehart and Company, 1955), B. M. Stewart, *Theory of Numbers* (New York, The Macmillan Company, 1952), W. J. LeVeque, *Topics in Number Theory*, Volume 1 (Reading,

Massachusetts, Addison-Wesley Publishing Company, 1956), and A. O. Gelfond, *The Solution of Equations in Integers*, translated from the Russian (San Francisco, W. H. Freeman and Co., 1961).

Example 5. Show how the number of steps involved in the Euclidean Algorithm can sometimes be reduced if both positive and negative remainders are permitted, and illustrate with the Example, § 407, and Example 3 above.

Solution. By finding the multiple of 364 that is *nearest* 9035, and similarly for subsequent divisions, we have:

$$(5) \quad \begin{cases} 9035 = 25 \cdot 364 - 65, \\ 364 = 6 \cdot 65 - 26, \\ 65 = 2 \cdot 26 + 13, \\ 26 = 2 \cdot 13. \end{cases}$$

Therefore, $(9035, 364) = 13$.

As in Example 3, $13 = 65 - 2 \cdot 26 = 65 - 2(6 \cdot 65 - 364) = 2 \cdot 364 - 11 \cdot 65 = 2 \cdot 364 - 11(25 \cdot 364 - 9035) = 9035 \cdot 11 - 364 \cdot 273$, with $m = 11$ and $n = 273$.

*409. EXERCISES

Unless specific information to the contrary is given, all numbers indicated in the following exercises are natural numbers.

In Exercises 1–6, find the GCD and the LCM of the two given numbers.

1. $a = 8, b = 12$.   **2.** $a = 35, b = 24$.

3. $a = 78, b = 42$.   **4.** $a = 560, b = 2250$.

5. $a = 1001, b = 111$.   **6.** $a = 420, b = 864$.

In Exercises 7–12, find natural numbers m, n, r, and s such that $ma - nb = rb - sa = (a, b)$, where a and b are the numbers given in the specified exercise.

7. Ex. 1.   **8.** Ex. 2.   **9.** Ex. 3.

10. Ex. 4.   **11.** Ex. 5.   **12.** Ex. 6.

13. Prove that the LCM of two natural numbers is their product if and only if they are relatively prime.

14. Find the GCD of 120 and 72 by reducing the fraction 120/72 to lowest terms. State and prove a general principle of which this is a particular illustration.

15. Prove that if $(a, b) = (a, c) = 1$, then $(a, bc) = 1$.

16. Use Theorem V, § 408, to construct a new proof of Theorem IV, § 404. *Hint:* Multiply both members of $ma - nc = 1$ by b.

17. Use Theorem V, § 408, to construct a new proof of Theorem III, § 408. (Cf. Ex. 16.)

18. If $(a, b) = 1$, $a \mid c$, and $b \mid c$, prove that $ab \mid c$.

19. Prove that natural numbers x and y exist such that $ax - by = c$, where a, b and c are given natural numbers, if and only if $(a, b) \mid c$.

20. Show that if a and b are any two given natural numbers, there exist prime numbers $p_1, p_2, \cdots, p_j$, and numbers $\alpha_1, \alpha_2, \cdots, \alpha_j$ and $\beta_1, \beta_2, \cdots, \beta_j$ each of which is either 0 or a natural number, such that

(1)
$$\begin{cases} a = p_1^{\alpha_1} p_2^{\alpha_2} \cdots p_j^{\alpha_j}, \\ b = p_1^{\beta_1} p_2^{\beta_2} \cdots p_j^{\beta_j}. \end{cases}$$

(Cf. Ex. 4, § 308.)   Prove that $a = b$ if and only if $\alpha_i = \beta_i$ for $i = 1, 2, \cdots, j$, that $a \mid b$ if and only if $\alpha_i \leq \beta_i$ for $i = 1, 2, \cdots, j$, and that

(2)
$$\begin{cases} ab = p_1^{\alpha_1 + \beta_1} p_2^{\alpha_2 + \beta_2} \cdots p_j^{\alpha_j + \beta_j}, \\ (a, b) = p_1^{\min(\alpha_1, \beta_1)} p_2^{\min(\alpha_2, \beta_2)} \cdots p_j^{\min(\alpha_j, \beta_j)}, \\ [a, b] = p_1^{\max(\alpha_1, \beta_1)} p_2^{\max(\alpha_2, \beta_2)} \cdots p_j^{\max(\alpha_j, \beta_j)}. \end{cases}$$

Prove that a and b are relatively prime if and only if $\alpha_1 \beta_1 = \alpha_2 \beta_2 = \cdots = \alpha_j \beta_j = 0$. Use (2) to give a new proof of Theorem IV, § 408.

21. Show that the representation (1) in terms of powers of a common set of prime numbers $p_1, p_2, \cdots, p_j$ can be extended to any given finite set $a, b, c, \cdots$ of natural numbers.   In particular, if a, b, and c are any three natural numbers, let them be expressed in the form (1) and

(3)
$$c = p_1^{\gamma_1} p_2^{\gamma_2} \cdots p_j^{\gamma_j}.$$

Use (1), and (2), and (3) to prove that $(ac, bc) = c(a, b)$.

22. Use Exercises 20 and 21 to construct new proofs of Theorem III, § 408, and of Exercises 15 and 18.

23. Prove that if $(b, c) = 1$, then $(a, bc) = (a, b)(a, c)$.   Cf. Exs. 20, 21, 24.) ·

24. Prove that $(a, [b, c]) = [(a, b), (a, c)]$ and $[a, (b, c)] = ([a, b], [a, c])$.   These are called **distributive laws**.   For a discussion of these and related matters, and their relation to the subject of *lattice theory*, see Oystein Ore, *Number Theory and Its History* (New York, McGraw-Hill Book Co., Inc., 1948), pp. 100–109.

25. Define the concepts of GCD and LCM of three natural numbers a, b, and c, denoted (a, b, c) and $[a, b, c]$, respectively.   With the notation of (1) and (3), show that

(4)
$$\begin{cases} (a, b, c) = p_1^{\min(\alpha_1, \beta_1, \gamma_1)} \cdots p_j^{\min(\alpha_j, \beta_j, \gamma_j)} \\ [a, b, c] = p_1^{\max(\alpha_1, \beta_1, \gamma_1)} \cdots p_j^{\max(\alpha_j, \beta_j, \gamma_j)}. \end{cases}$$

Establish the associative laws:

(5)
$$\begin{cases} (a, b, c) = ((a, b), c) = ((a, c), b) = (a, (b, c)), \\ [a, b, c] = [[a, b], c] = [[a, c], b] = [a, [b, c]]. \end{cases}$$

State and prove corresponding commutative laws.

26. Use the representations of Exercise 25 to prove that $(ad, bd, cd) = d(a, b, c)$. State and prove distributive laws for three numbers similar to those of Exercise 24.

27. In the notation of Exercise 25, prove that

$$([a, b], [a, c], [b, c]) = [(a, b), (a, c), (b, c)].$$

28. Extend the concepts of GCD and LCM to an arbitrary finite set of natural numbers.   Formulate and prove a few general theorems of your own design.

5

Integers and Rational Numbers

+++

501. THE SYSTEM OF INTEGERS

Let $\mathscr{G}$ denote an arbitrary ordered field (which may, in particular, be the set of all real numbers). Then the subset of $\mathscr{G}$ consisting of those members that are known as *integers* or *whole numbers* is defined in terms of the set $\mathscr{N}$ of natural numbers. We continue to call the members of $\mathscr{G}$ *numbers*.

Definition. *A number x is an **integer** if and only if (i) $x \in \mathscr{N}$, (ii) $x = 0$, or (iii) $-x \in \mathscr{N}$. The set of all integers is denoted $\mathscr{I}$.*

Theorem I. *If x is an integer, then exactly one of the three statements listed in the preceding definition is true. A number is a positive integer if and only if it is a natural number (statement (i)). A number is a negative integer if and only if its negative is a positive integer, or, equivalently, a natural number (statement (iii)).*

Proof. Since $\mathscr{N}$ is a subset of the set $\mathscr{P}$ of positive numbers, at most one of the three alternatives is possible, by Axioms IV, § 202. On the other hand, by the definition of an integer, for any integer x at *least* one of the three must hold. If an integer x is positive, alternatives (ii) and (iii) are eliminated and x must belong to $\mathscr{N}$. Conversely, every member of $\mathscr{N}$ is positive and, by definition, an integer. In a similar fashion we see that a number x is a negative integer if and only if alternative (iii) is true.

Theorem II. *The system of integers $\mathscr{I}$ is closed with respect to addition, multiplication, and subtraction, but not with respect to division. That is, if m and n are integers, then $m + n$, mn, and $m - n$ are integers, but there exist integers m and n such that m/n is not an integer (even when defined).*

Proof. We start with addition. There are several cases. In the first place, if either m or n is 0 the result is obvious. If both m and n are positive

62

the result follows from the fact that the system of natural numbers is closed with respect to addition (Property III, § 304). If both m and n are negative, then $m + n = -[(-m) + (-n)]$, and this quantity is the negative of the natural number in brackets, and hence an integer. Finally, we shall assume that m and n have opposite signs, and adjust the notation, if necessary, so that m is a positive integer and n a negative integer; that is, m and $-n$ are both natural numbers. There are three subcases remaining. First: If $m > -n$, then $m + n = m - (-n)$, which is a natural number by Property V, § 304. Second: If $m = -n$, then $m + n = m - (-n) = 0$. Third: If $m < -n$, then $m + n = -[(-n) - m]$, which is an integer since the bracketed quantity is a natural number by Property V, § 304.

Multiplication is easier. If either m or n is 0 so is their product. If both are positive their product is a natural number by Property IV, § 304. If both are negative their product mn is again a natural number since $mn = (-m)(-n)$. If m and n have opposite signs, say $m > 0$ and $n < 0$, then mn is the negative of the natural number $m(-n)$.

Since $-n$ is an integer if and only if n is an integer, $m - n = m + (-n)$ is an integer whenever m and n are integers.

Finally, as shown in the Note, § 304, the number 1/2 is not a natural number. Therefore 1/2 is not an integer, since if it were an integer it would be a positive integer and hence a natural number.

Note. If an algebraic system S with two binary operations, called *addition* and *multiplication*, subject to all of the field axioms I, II, and III, § 102, except that the existence of a reciprocal (Axiom II (iv), § 102) is replaced by the weaker assumption of Example 4, § 103, that the product of any two nonzero members of S is nonzero, then S is called an **integral domain**. By Theorem II *the integers form an integral domain*. (For another example of an integral domain that is not a field, cf. § 702.) An important property of any integral domain is that *the cancellation law for multiplication holds:* $xy = xz$ implies $y = z$ if $x \neq 0$. (This is true in any integral domain since it follows from the property of Example 4, § 103, thus: $xy = xz$ implies $xy - xz = x(y - z) = 0$; if the product of nonzero members of S is always nonzero, then not both factors, x and $y - z$, can be nonzero, and therefore if x is nonzero, $y - z$ must be equal to 0, or $y = z$.) By Theorem I, § 403, the integral domain of the integers is **Archimedean**: *If a and b are positive integers there exists a positive integer n such that $na > b$.*

502. A NATURAL ISOMORPHISM

In § 311 it was shown that any two natural number systems in any two ordered fields are abstractly identical in the sense that there is an addition-preserving, multiplication-preserving, and order-preserving one-to-one correspondence, or *isomorphism*, between the members of these two natural number systems. The definition of isomorphism given in § 311 for natural number systems applies without change to systems of integers in arbitrary ordered fields. Of considerable importance and interest is the fact that the Theorem, § 311, which states that any two natural number systems are isomorphic, applies in equal force to any two systems of integers:

Theorem. *Any two systems of integers are isomorphic. That is, if $\mathscr{I}$ is the system of integers of an ordered field $\mathscr{G}$ and if $\mathscr{I}'$ is the system of integers of an ordered field $\mathscr{G}'$, then $\mathscr{I}$ and $\mathscr{I}'$ are isomorphic.*

Proof. The correspondence $n \leftrightarrow n'$, where n is an arbitrary integer of $\mathscr{I}$, is defined in three parts: (*i*) if $n > 0$, then n is a natural number, and n' is the natural number of $\mathscr{I}'$ that corresponds to n by the isomorphism of § 311; (*ii*) if $n = 0$ then $n' \equiv 0'$ is the zero of $\mathscr{G}'$; (*iii*) if $n < 0$ then $n' \equiv -(-n)'$. The proof that the correspondence $n \leftrightarrow n'$ is one-to-one is easy, and will be left to the reader to supply.

The fact that this correspondence is addition-preserving resolves itself into several cases, as in the proof of Theorem II, § 501. The problem is to show that if m and n are integers, then $(m + n)' = m' + n'$. In the first place, if either m or n is 0 the result is obvious. If both m and n are positive the result follows from the addition-preserving property of the isomorphism already established for the natural numbers in § 311. If both m and n are negative, then $(m + n)' = -[-(m + n)]' = -[(-m) + (-n)]' = -[(-m)' + (-n)'] = [-(-m)'] + [-(-n)'] = m' + n'$. Finally, for addition-preserving, we shall assume that m and n have opposite signs, and adjust the notation, if necessary, so that m is a positive integer and n a negative integer. There are three subcases remaining. First: If $m > -n$, then $m + n = m - (-n) > 0$, and hence, by the Corollary, § 311, which establishes the fact that the isomorphism of natural numbers preserves subtraction whenever the difference is a natural number, $(m + n)' = [m - (-n)]' = m' - (-n)' = m' + n'$. Second: If $m = -n$, then $m' = (-n)' = -n'$, so that $(m + n)' = 0' = m' + n'$. Third: If $m < -n$, then $(m + n)' = -[-(m + n)]' = -[(-n) - m]' = -[(-n)' - m'] = -[-n' - m'] = m' + n'$.

Proof that the correspondence $n \leftrightarrow n'$ is multiplication-preserving is easier. If either m or n is 0 so is their product, and $(mn)' = 0' = m'n'$. If both m and n are positive, then $(mn)' = m'n'$ by the isomorphism of § 311. If both are negative, then $(mn)' = [(-m)(-n)]' = (-m)'(-n)' = [-(m')][-(n')] = m'n'$. If m and n have opposite signs, say $m > 0$ and $n < 0$, then $(mn)' = -[-(mn)]' = -[m(-n)]' = -[m'(-n)'] = -[m'(-n')] = m'n'$.

Finally, $m < n$ if and only if there exists a positive integer p such that $m + p = n$, and this relation is true if and only if $(m + p)' = m' + p' = n'$; that is (since $p' > 0$ if and only if $p > 0$), if and only if $m' < n'$. Therefore the isomorphism under consideration is order-preserving, and the proof is complete.

Corollary. *Any ordered field contains an integral domain of integers isomorphic to the integral domain of integers of the real number system.*

503. LAWS OF EXPONENTS

The laws of exponents enumerated in § 307 find ready extension to cases where the exponents are arbitrary integers, rather than being restricted to being positive integers. (Cf. Ex. 4, § 308.) First we give a definition.

Definition. *If x is any real number†* (*or, more generally, any member of an arbitrary field*), *then* $x^0 \equiv 1$. *If* $x \neq 0$ *and n is a negative integer, then* $x^n \equiv 1/x^{-n}$.

Examples. $3^0 = 1$, $2^{-1} = \frac{1}{2}$, $(\frac{2}{3})^{-4} = \frac{81}{16}$.

Theorem. *If x and y are any nonzero numbers* (*or, more generally, nonzero members of an arbitrary field*) *and if m and n are any integers, then:*

(*i*) $$x^m x^n = x^{m+n};$$

(*ii*) $$\frac{x^m}{x^n} = x^{m-n};$$

(*iii*) $$(x^m)^n = x^{mn};$$

(*iv*) $$(xy)^n = x^n y^n;$$

(*v*) $$\left(\frac{x}{y}\right)^n = \frac{x^n}{y^n}.$$

Proof. (*i*): If either *m* or *n* (or both) is equal to 0 the result is obvious. If both *m* and *n* are positive the result follows from part (*i*) of the Theorem, § 307. If both *m* and *n* are negative, let $p \equiv -m$ and $q \equiv -n$. Then *p* and *q* are positive integers and (*i*) becomes $(1/x^p)(1/x^q) = 1/x^{p+q}$, which is true, again by part (*i*) of the Theorem, § 307, and elementary properties of numbers from Chapter 1. Finally, we shall assume that *m* and *n* have opposite signs, and adjust the notation, if necessary, so that $m > 0$ and $n < 0$. Letting $q \equiv -n$, we have the task of verifying the equation $x^m/x^q = x^{m-q}$. If $m > q$ or if $m < q$ this is true by part (*ii*) of the Theorem, § 307 (in the second of these two cases $x^{m-q} = 1/x^{q-m}$). If $m = q$, $x^m/x^q = 1 = x^{m-q}$, and the proof of (*i*) is complete.

(*ii*): By (*i*), $x^m = x^{m-n}x^n$, and division by x^n gives (*ii*).

(*iii*): If either *m* or *n* (or both) is equal to 0, both members of (*iii*) are equal to 1. If both *m* and *n* are positive, the result follows from part (*iii*) of the Theorem, § 307. If both *m* and *n* are negative, let $p \equiv -m$ and $q \equiv -n$. Then the problem is to establish the equality $1/(1/x^p)^q = x^{pq}$. By parts (*v*) and (*iii*) of the Theorem, § 307, the left-hand member of the present equation

† For present purposes we shall define 0^0 to be equal to 1, although in calculus there are good reasons for considering 0^0 to be *indeterminate*, and restricting the definition of x^0 to *nonzero* values of *x*. Our principal reason for including $x = 0$ in the definition of $x^0 = 1$ is that whenever the quantities involved are defined, even with base 0, the standard laws of exponents apply. (Cf. Ex. 4, § 308.)

is equal to $1/(1^q/(x^p)^q) = 1/(1/x^{pq}) = x^{pq}$. If $m > 0$ and $n < 0$, let $q \equiv -n$. Then (iii) reduces to $1/(x^m)^q = 1/x^{mq}$, which is true by § 307. Finally, if $m < 0$ and $n > 0$, let $p \equiv -m$. Then (iii) reduces to $(1/x^p)^n = 1/x^{pn}$, which is again true by § 307.

(iv): If $n > 0$, this is part of the Theorem, § 307. If $n = 0$, both members of (iv) are equal to 1. If $n < 0$, let $q \equiv -n$. Then (iv) reduces to $1/(xy)^q = (1/x^q)(1/y^q)$, which is true by § 307.

(v): By (iv), $y^n \left(\dfrac{x}{y}\right)^n = \left(y\,\dfrac{x}{y}\right)^n = x^n$, and division by y^n gives (v).

504. THE ORDERED FIELD OF RATIONAL NUMBERS

Definition. *A number x is a **rational number** if and only if there exist integers m and n, where $n \neq 0$, such that $x = m/n$. The real numbers that are not rational are called **irrational**. The set of all rational numbers is denoted $\mathscr{Q}$.*

Theorem. *The system $\mathscr{Q}$ of rational numbers is an ordered field.*

Proof. We start by examining in turn the axioms of § 102. Axiom I(i) takes the form: *the sum of two rational numbers is rational.* This follows from the rule for adding fractions (Ex. 16, § 104): $\dfrac{m}{n} + \dfrac{p}{q} = \dfrac{mq + np}{nq}$ (where $n \neq 0$ and $q \neq 0$). Since m, n, p, and q are integers, the numerator and the denominator of the resulting fraction are both integers (§ 502), and the denominator nq is nonzero (Example 4, § 103). Axioms I(ii) and I(v) are true for all rational numbers since they are true for all real numbers. Axioms I(iii) and I(iv) hold since 0 is a rational number $(0 = 0/1)$, and if x is a rational number m/n then so is $-x = (-m)/n$. Axioms II and III follow similarly. II(i): $\dfrac{m}{n} \cdot \dfrac{p}{q} = \dfrac{mp}{nq}$. II(ii) and II(v) and III: These are true for all rational numbers since they are true for all real numbers. II(iii): $1 = 1/1$. II(iv): If $x = m/n$, where $n \neq 0$, is nonzero then $m \neq 0$ (Example 7, § 103), and $x^{-1} = n/m$.

To see that Axioms IV (§ 202) are satisfied for the system $\mathscr{Q}$ we take as the subset $\mathscr{P}$ all those rational numbers that are positive. Then IV(i) and IV(ii) follow immediately. Furthermore, if x is an arbitrary rational number, either x is positive (and rational) or x is zero (and rational) or $-x$ is positive (and rational). Therefore IV(iii) is also satisfied, and the rational numbers form an ordered field.

NOTE. We have the following inclusion relations among the system $\mathscr{N}$ of natural numbers, the integral domain $\mathscr{I}$ of integers, the ordered field $\mathscr{Q}$ of rational numbers, and the ordered field $\mathscr{R}$ of real numbers:

$$\mathscr{N} \subset \mathscr{I} \subset \mathscr{Q} \subset \mathscr{R}.$$

Each of these inclusions is a proper inclusion. That is, there are integers that are not natural numbers (-1 is an example), there are rational numbers that are not integers ($\frac{1}{2}$ is an example), and there are real numbers that are not rational. At this point it is impossible to establish the existence of irrational numbers (that is, real numbers that are not rational), since the rational number system satisfies all of the axioms that are at our disposal. In § 905, with the aid of the Axiom of Completeness, we shall exhibit examples of irrational numbers.

505. A NATURAL ISOMORPHISM

It has already been established, in §§ 311 and 502, that in any two ordered fields any two natural number systems are isomorphic and that any two systems of integers are isomorphic, with respect to addition, multiplication, and order. In the present section we show that this "natural isomorphism" between basic systems within any two ordered fields extends to the rational numbers:

Theorem. *Any two rational number systems are isomorphic. That is, if $\mathcal{Q}$ is the system of rational numbers of an ordered field $\mathcal{G}$ and if $\mathcal{Q}'$ is the system of rational numbers of an ordered field $\mathcal{G}'$, then $\mathcal{Q}$ and $\mathcal{Q}'$ are isomorphic.*

Proof. If x is an arbitrary member of $\mathcal{Q}$ then x', as a member of $\mathcal{Q}'$, is defined as follows: If p and q are integers such that $x = p/q$, let p' and q' be the integers of $\mathcal{G}'$ that correspond to p and q, respectively, according to the natural isomorphism between $\mathcal{I}$ and $\mathcal{I}'$ described in § 502. Then x' is defined to be p'/q'. In other words, $(p/q)' = p'/q'$, and the one-to-one correspondence studied in §§ 311 and 502 extends by the rule:

$$\frac{p}{q} \leftrightarrow \frac{p'}{q'} .$$

The first item to check is that there is no ambiguity in this rule of correspondence; that is, that if $p/q = r/s$, then $p'/q' = r'/s'$. This follows (§ 502) from the fact that if $ps = qr$, then $p's' = q'r'$. It remains to show that

$$\left(\frac{p}{q} + \frac{u}{v}\right)' = \frac{p'}{q'} + \frac{u'}{v'} \quad \text{and} \quad \left(\frac{p}{q} \cdot \frac{u}{v}\right)' = \frac{p'}{q'} \cdot \frac{u'}{v'} ,$$

and that

$$\frac{p}{q} < \frac{u}{v} \quad \text{implies} \quad \frac{p'}{q'} < \frac{u'}{v'} .$$

The first two are simple:

$$\left(\frac{p}{q} + \frac{u}{v}\right)' = \left(\frac{pv + qu}{qv}\right)' = \frac{(pv + qu)'}{(qv)'} = \frac{p'v' + q'u'}{q'v'} = \frac{p'}{q'} + \frac{u'}{v'} ,$$

and

$$\left(\frac{p}{q} \cdot \frac{u}{v}\right)' = \left(\frac{pu}{qv}\right)' = \frac{(pu)'}{(qv)'} = \frac{p'u'}{q'v'} = \frac{p'}{q'} \cdot \frac{u'}{v'} .$$

The order preservation is based on the fact that positive elements correspond to positive elements. If m/n is positive (in $\mathscr{G}$), then m and n have the same sign and hence m' and n' have the same sign (in $\mathscr{G}'$), and m'/n' is positive. Consequently, if p/q is less than u/v, then u/v is equal to p/q plus a positive fraction r/s. Therefore u'/v' is equal to p'/q' plus a positive fraction r'/s', and hence p'/q' is less than u'/v'.

Corollary. *Any ordered field contains an ordered field of rational numbers isomorphic to the ordered field of rational numbers of the real number system.*

506. EXERCISES

In the following exercises, unless explicit exception is made, all quantities under consideration—referred to for simplicity as *numbers*—are arbitrary members of an ordered field $\mathscr{G}$.

1. If 1 is the unity of an arbitrary field, and if n is any integer, prove that $1^n = 1$.

2. Prove that the sum of any two negative integers is a negative integer, and that the product of any two negative integers is a positive integer.

3. Prove that the sum of any finite number of negative integers is a negative integer.

4. If $a > 1$ and if m and n are arbitrary integers, prove that the inequality $m < n$ is equivalent to the inequality $a^m < a^n$. If $0 < a < 1$ and if m and n are arbitrary integers, prove that the inequality $m < n$ is equivalent to the inequality $a^m > a^n$. (Cf. Ex. 6, § 308.)

5. If $a > 0$, and $a \neq 1$, and if m and n are integers, prove that $a^m = a^n$ if and only if $m = n$. (Cf. Ex. 7, § 308, Ex. 4 above.)

6. If a and b are positive members of an ordered field and if n is a nonzero integer, prove that $a^n = b^n$ if and only if $a = b$. (Cf. Ex. 8, § 308.)

7. If a and b are nonzero members of an ordered field and if n is any integer prove that $a^{2n+1} = b^{2n+1}$ if and only if $a = b$. (Cf. Ex. 9, § 308.)

8. An ordered field $\mathscr{G}$ is called **Archimedean** if and only if for any two positive members a and b of $\mathscr{G}$ there exists a natural number n such that $na > b$. Prove that the ordered field $\mathscr{Q}$ of rational numbers is Archimedean. (Cf. §§ 403, 501, 713, 903.)

9. Prove that there do not exist two integers m and n such that $m < n < m + 1$.

10. Prove that between any two integers there exist only finitely many integers.

11. Let S be a nonempty set of integers such that there exists an integer N greater than every member of S. Prove that S has a greatest member.

12. Prove the following variant of the Fundamental Theorem of Mathematical Induction (Theorem III, § 303): *Let N be an arbitrary but fixed integer, and for every integer $n \geq N$ let $P(n)$ be a proposition that is either true or false. If* (i) *$P(N)$ is true and* (ii) *whenever the proposition $P(n)$ is true for the integer $n \geq N$ the proposition $P(n + 1)$ is also true, then $P(n)$ is true for every integer $n \geq N$.* Hint: Let $Q(n) \equiv P(N + n - 1)$. (Cf. Ex. 33, § 308.)

13. If n is an integer and if m is a positive integer, establish the unique existence of an integer q and a nonnegative integer r such that $r < m$ and $n = qm + r$.

6

*Congruences and Finite Fields

*601. INTRODUCTION

In Chapter 3 (§ 309) a relation from a nonempty set A to a nonempty set B was defined as any subset of the Cartesian product $A \times B$. The emphasis in that chapter was on the particular relations known as *functions*. We turn our attention now to another kind of relation, known as a *transitive relation*, with special attention to its particularization known as an *equivalence relation*. The symbol $\mathscr{G}$ denotes an arbitrary ordered field—which may, in particular, be chosen to be the real number system $\mathscr{R}$—$\mathscr{N}$ denotes the system of natural numbers of $\mathscr{G}$, $\mathscr{I}$ denotes the integral domain of integers of $\mathscr{G}$, and $\mathscr{Q}$ denotes the ordered field of rational numbers of $\mathscr{G}$.

*602. TRANSITIVITY

Let A be a nonempty set and let ρ be a relation from A to A—that is, ρ is a subset of the Cartesian product $A \times A$. Then ρ is **transitive** if and only if $x \,\rho\, y$ and $y \,\rho\, z$ imply $x \,\rho\, z$.

Examples 1. Each of the following defines a transitive relation from $\mathscr{R}$ to $\mathscr{R}$ (the student should supply each verification): (*i*) $x < y$; that is, the subset of the Cartesian plane consisting of all points (x, y) such that $x < y$. (*ii*) $x \leq y$; that is, the subset of the Cartesian plane consisting of all points (x, y) such that $x \leq y$. (*iii*) $x > y$; that is, the subset of the Cartesian plane consisting of all points (x, y) such that $x > y$. (*iv*) $x \geq y$; that is, the subset of the Cartesian plane consisting of all points (x, y) such that $x \geq y$. (*v*) $x = y$; that is, the subset of the Cartesian plane consisting of all points with equal coordinates.† (*vi*) $x \,\rho\, y$ if and only if $x - y \in \mathscr{N}$. (*vii*) $x \,\rho\, y$ if and only if $x - y \in \mathscr{I}$. (*viii*) $x \,\rho\, y$ if and only if $x - y \in \mathscr{Q}$. (*ix*) $x \,\rho\, y$ if and only if $x \in \mathscr{N}, y \in \mathscr{N}$, and $x - y \in \mathscr{N}$. (*x*) $x \,\rho\, y$ if and only if x and y are both integers. (*xi*) $x \,\rho\, y$ is universally true; that is, $\rho = \mathscr{R} \times \mathscr{R}$. (*xii*) $x \,\rho\, y$ is universally false; that is, $\rho = \varnothing$.

---

* Starred sections and chapters may be omitted without destroying the continuity of the remaining material. See the preface for a discussion of the role of starring in this book.

† More generally, the **identity relation** from any nonempty set to itself, consisting of all ordered pairs that have equal coordinates, is transitive.

Examples 2. Each of the following relations from $\mathcal{R}$ to $\mathcal{R}$ fails to be transitive: (*i*) $x \neq y$; that is, the subset of the Cartesian plane consisting of all points with unequal coordinates (for example, if ρ is the relation of being unequal, if $x = 0$, $y = 1$, and $z = 0$, then $x \rho y$ is true and $y \rho z$ is true, but $x \rho z$ is false). (*ii*) $x \rho y$ if and only if $x + y \in \mathscr{I}$ (for example, if $x = \frac{1}{3}$, $y = \frac{2}{3}$, and $z = \frac{1}{3}$, then $x \rho y$ is true and $y \rho z$ is true, but $x \rho z$ is false). (*iii*) $x \rho y$ if and only if $y = f(x)$, where f is a given function on $\mathcal{R}$ into $\mathcal{R}$, unless f is **idempotent**: $f(f(x)) = f(x)$ for every x in $\mathcal{R}$ (an example of an idempotent function is the greatest integer function discussed in the Note, § 903).

*603. EQUIVALENCE RELATIONS

Let A be a nonempty set and let ρ be a relation from A to A. Then ρ is **reflexive** if and only if $x \rho x$ for every $x \in A$. The relation ρ is **symmetric** if and only if $x \rho y$ implies $y \rho x$ (that is, whenever $x \rho y$ is true, $y \rho x$ is also true). An **equivalence relation** from A to A is a relation, commonly written $\sim$, that is reflexive, symmetric, and transitive. The statement $x \sim y$ is also phrased x *is equivalent to* y or, because of symmetry (since x is equivalent to y if and only if y is equivalent to x), x *and* y *are equivalent.*

Examples 1. The relations of Examples 1, § 602, that are equivalence relations are (*v*),† (*vii*), (*viii*), and (*xi*), as the reader may easily verify. In examples (*i*), (*ii*), (*iii*), (*iv*), (*vi*), and (*ix*) ρ is not symmetric (take $x = 1$ and $y = 2$, or $x = 2$ and $y = 1$). In examples (*i*), (*iii*), (*vi*), (*ix*), (*x*), and (*xii*) ρ is not reflexive since $\frac{1}{2} \rho \frac{1}{2}$ is false in each instance.

Closely allied to the concept of equivalence relation is that of *equivalence class.*‡ If $\sim$ is an equivalence relation from a nonempty set A to itself, then the **equivalence class of a member** x of A is the set of all members y of A that are equivalent to x: $y \sim x$. In general, an **equivalence class** is the equivalence class of some member of A. By reflexivity, the equivalence class of any member x of A always contains x.

The following theorem shows that an equivalence class is equally well determined by any one of its members.

Theorem I. *If $\sim$ is an equivalence relation from a nonempty set A to itself, any equivalence class is the equivalence class of any one of its members. In other words, if X is the equivalence class of x, consisting of all members y of A that are equivalent to x, and if z is an arbitrary member of X, then the equivalence class Z of z, consisting of all members u of A that are equivalent to z, is identical with X: $Z = X$.*

Proof. We first show that Z is a subset of X: $Z \subset X$. Accordingly, let u be an arbitrary member of Z: $u \in Z$. By definition of Z, $u \sim z$, and by

---

† More generally, the identity relation from any nonempty set to itself is an equivalence relation.

‡ The word *class*, and the word *family* introduced below, are used occasionally in place of the word *set*, for the sake of euphony and variety. The meaning is the same.

definition of X, $z \sim x$. Therefore, by transitivity, $u \sim x$, and again by definition of X, $u \in X$, and we have the desired inclusion, $Z \subset X$. In a similar way we can show that X is a subset of Z: If $u \in X$, then $u \sim x$. By symmetry, since $z \sim x$, $x \sim z$, and thus by transitivity $u \sim z$, and $u \in Z$. Since $Z \subset X$ and $X \subset Z$ we conclude that $Z = X$, as desired.

Intimately associated with the concept of an equivalence relation from a nonempty set A to itself is that of *partition*. A **partition** of A is a family of nonempty subsets X, Y, $\cdots$ of A such that (*i*) every x in A is a member of some subset X of the partition and (*ii*) no member of A is a member of two distinct subsets of the partition. In other words, the subsets of A that form a partition are nonempty, exhaust A, and are pairwise disjoint. The precise connection between equivalence relations and partitions is stated in the theorem:

Theorem II. *If $\sim$ is an equivalence relation from a nonempty set A to itself, then the family of all equivalence classes of A is a partition of A. Conversely, if Π is any partition of A, and if $x \sim y$ is defined to mean that x and y belong to the same subset of A in the partition Π, then $\sim$ is an equivalence relation, and the family of equivalence classes of $\sim$ is the partition Π.*

Proof. If $\sim$ is an equivalence relation from A to A, and if x is an arbitrary number of A, then the equivalence class of x contains x, and hence the family of all equivalence classes exhausts A. To show that distinct equivalence classes are disjoint, assume that X and Y are equivalence classes that are not identical but have an element z in common. By Theorem I, if Z is the equivalence class of z, then $Z = X$ and $Z = Y$. Therefore $X = Y$, in contradiction to the assumption that X and Y are not identical. Conversely, if the assumptions of the second sentence of the theorem are made, it is trivial to verify that $\sim$ is an equivalence relation. An equivalence class of any x according to $\sim$ is the set of all y that belong to the same set of the partition Π as x does, and is therefore one of the sets of Π, whereas every set of the partition Π is simply the equivalence class of any one of its members.

Examples 2. If $\sim$ is the equivalence relation of (*v*), Examples 1, § 602, that is, $x \sim y$ if and only if $x = y$, then every equivalence class consists of exactly one number. If $\sim$ is the equivalence relation of (*vii*), Examples 1, § 602, that is, if $x \sim y$ if and only if $x - y$ is an integer, then the equivalence class of x consists of all numbers of the form $n + x$, where n is an integer. Every equivalence class is uniquely determined by some x such that $0 \leq x < 1$ (cf. § 903). If $\sim$ is the equivalence relation of (*viii*), Examples 1, § 602, that is, $x \sim y$ if and only if $x - y$ is rational, then the equivalence class of x consists of all numbers of the form $r + x$, where r is a rational number. If $\sim$ is the equivalence relation of (*xi*), Examples 1, § 602, that is, if $x \sim y$ is universally true, there is only one equivalence class, $\mathscr{R}$ itself.

*604. CONGRUENCES

For present and future purposes it will be useful to extend the concept and notation of divisibility introduced in § 401 for natural numbers, so that they apply to all integers, as follows: A nonzero integer m **divides** an integer k (in symbols: $m \mid k$) if and only if there exists an integer n such that $k = mn$. If $m \mid k$, m is called a **divisor** or **factor** of k, and k is called a **multiple** of m. In particular, every nonzero integer m is a divisor of 0: $m \mid 0$. Two nonzero integers m and n are **relatively prime** if and only if the only natural number that divides both m and n is 1.

Let m be a fixed natural number, and for two integers a and b let a relation ρ be defined: $a \, \rho \, b$ if and only if $m \mid (a - b)$. It follows directly from this definition that ρ is an equivalence relation from $\mathscr{I}$ to $\mathscr{I}$ (this is true for the following three reasons: $m \mid (a - a)$ for every a; if $m \mid (a - b)$ then $m \mid (b - a)$; if $m \mid (a - b)$ and $m \mid (b - c)$ then $m \mid (a - c)$ since $a - c = (a - b) + (b - c)$).

The German mathematician Carl Friedrich Gauss (1777–1855) introduced the following notation for the equivalence relation just defined:

(1) $a \equiv b \pmod{m}$ if and only if $m \mid (a - b)$.

The relation (1) is stated in words: *a is congruent to b modulo m*. The number m is called the **modulus** of the **congruence** (1). The statement that (1) is false is written:

(2) $a \not\equiv b \pmod{m}$;

in words: *a is incongruent to b modulo m*.

The congruence notation (1) resembles an equation. One reason for the appropriateness of that notation is that congruences, in their elementary behavior, are similar to equations.

Theorem I. *If a, b, c, d, and n are integers, if m is a natural number, and if $a \equiv b \pmod{m}$ and $c \equiv d \pmod{m}$, then $a + c \equiv b + d \pmod{m}$, and $a - c \equiv b - d \pmod{m}$, $ac \equiv bd \pmod{m}$, and $na \equiv nb \pmod{m}$.*

Proof. Assume that $m \mid (a - b)$ and $m \mid (c - d)$. Then, by elementary principles, m is a divisor of the sum $(a - b) + (c - d) = (a + c) - (b + d)$, so that $a + c \equiv b + d \pmod{m}$. Similarly, m is a divisor of the difference $(a - b) - (c - d) = (a - c) - (b - d)$, so that $a - c \equiv b - d \pmod{m}$. If we rewrite $ac - bd$ by adding and subtracting bc: $ac - bd = ac - bc + bc - bd = (a - b)c + b(c - d)$, we see that since m divides both $(a - b)$ and $(c - d)$, m divides the expanded form of $ac - bd$, so that $ac \equiv bd \pmod{m}$. Finally, since $na - nb = n(a - b)$, $m \mid (na - nb)$ and $na \equiv nb \pmod{m}$.

The law of cancellation does not in general hold. That is, it is possible for the congruence $na \equiv nb \pmod{m}$ to hold and for the congruence $a \equiv b \pmod{m}$ to fail, even when $n \not\equiv 0 \pmod{m}$. For example, if $a = 5$, $b = 2$, $n = 4$, and $m = 6$, we have $4 \cdot 5 \equiv 4 \cdot 2 \pmod 6$, whereas both $5 \not\equiv 2 \pmod 6$ and $4 \not\equiv 0 \pmod 6$. However, the cancellation law does hold in an important special case:

Theorem II. *If a, b, and n are integers, with $n \neq 0$, if m is a natural number such that m and n are relatively prime, and if $na \equiv nb \pmod{m}$, then $a \equiv b \pmod{m}$.*

Proof. Assume that $na \equiv nb \pmod{m}$. If $n > 0$ and $a \neq b$, adjust the notation for a and b, if necessary, so that $a > b$. Then, by assumption, $m \mid (na - nb)$, or $m \mid n(a - b)$. By Theorem IV, § 404, since m and n are relatively prime, $m \mid (a - b)$, and $a \equiv b \pmod{m}$. Finally, assume that $n < 0$ and let $p \equiv -n > 0$. Once more, assume for definiteness that $a > b$. Then, by assumption, $m \mid (na - nb)$, and hence $m \mid p(a - b)$. Again, since m and p are relatively prime, $m \mid (a - b)$, and $a \equiv b \pmod{m}$. This completes the proof.

*605. OPERATIONS WITH EQUIVALENCE CLASSES

To obtain a clearer picture of the equivalence classes given by congruence modulo m, as defined in § 604, we enumerate their members:

$$(1) \begin{cases} \cdots & -2m, & -m, & 0, & m, & 2m, & \cdots \\ \cdots & 1 - 2m, & 1 - m, & 1, & 1 + m, & 1 + 2m, & \cdots \\ \cdots & 2 - 2m, & 2 - m, & 2, & 2 + m, & 2 + 2m, & \cdots \\ & & & \cdots & & & \\ \cdots (m-1) - 2m, & (m-1) - m, & m - 1, & (m-1) + m, & (m-1) + 2m, & \cdots. \end{cases}$$

From the form of (1) we see that there are exactly m equivalence classes, each with infinitely many members, and that these equivalence classes can be represented by the m integers:

$$(2) \qquad\qquad 0, 1, 2, \cdots, m - 1.$$

(Cf. Ex. 13, § 506.)

If the notation $[n]$, $0 \leq n \leq m - 1$, is used to represent the equivalence class of the integer n, we can write the m equivalence classes thus:

$$(3) \qquad\qquad [0], [1], [2], \cdots, [m - 1].$$

If we now concentrate on these equivalence classes in terms of congruence as an equivalence relation, we find a very interesting algebra arising from the manipulation of these equivalence classes. If A and C are any two equivalence classes (1) (or, equivalently, (3)) and if n is any integer we shall formulate a definition for adding, subtracting, and multiplying A and C, and for multiplying A by n. It turns out that our definition needs justification, to show that it is *meaningful*, and that this justification is merely a restatement of Theorem I, § 604.

Definition. *Let m be a fixed natural number, and let congruence modulo m be the equivalence relation defined in (1), § 604. If A and C are any two equivalence classes and if n is any integer, define the following operations in terms of an arbitrary member a chosen from A and an arbitrary member c chosen from C:*

(i) $A + C =$ the equivalence class of $a + c$,

(ii) $A - C =$ the equivalence class of $a - c$,

(iii) $AC =$ the equivalence class of ac,

(iv) $nA =$ the equivalence class of na.

Theorem. *The preceding definition is meaningful in the sense that each of the four operations (i)–(iv) is independent of the representatives a and c chosen.*

Proof. For (*i*), let *b* be an arbitrary member of *A*, so that $a \equiv b \pmod{m}$, and let *d* be an arbitrary member of *C*, so that $c \equiv d \pmod{m}$. Then, by Theorem I, § 604, since $a + c \equiv b + d \pmod{m}$, the equivalence class of $b + d$ is the same as that of (*i*). For (*ii*), (*iii*), and (*iv*) the details follow in like manner from their counterparts in Theorem I, § 604.

Example 1. Write out the addition and multiplication tables for the equivalence classes for congruence modulo 4.

Solution. With the notation (3), these tables are:

	[0]	[1]	[2]	[3]
[0]	[0]	[1]	[2]	[3]
[1]	[1]	[2]	[3]	[0]
[2]	[2]	[3]	[0]	[1]
[3]	[3]	[0]	[1]	[2]

	[0]	[1]	[2]	[3]
[0]	[0]	[0]	[0]	[0]
[1]	[0]	[1]	[2]	[3]
[2]	[0]	[2]	[0]	[2]
[3]	[0]	[3]	[2]	[1]

Addition   *Multiplication*

In practice it is often convenient to remove the brackets in the designation of the equivalence classes [*n*], $1 \leq n \leq m - 1$, if it is clearly understood that when this is done *the symbol n represents a member of a new algebraic structure and is no longer a member of the real number system.* In spite of this we shall informally refer to the algebra of equivalence classes modulo *m* as that of *integers reduced modulo m*, or simply the *integers modulo m*.

Example 2. Write out the addition and multiplication tables for the integers modulo 5.

Solution. With the notation *n* for [*n*], these tables are:

	0	1	2	3	4
0	0	1	2	3	4
1	1	2	3	4	0
2	2	3	4	0	1
3	3	4	0	1	2
4	4	0	1	2	3

	0	1	2	3	4
0	0	0	0	0	0
1	0	1	2	3	4
2	0	2	4	1	3
3	0	3	1	4	2
4	0	4	3	2	1

Addition   *Multiplication*

*606. RINGS AND FIELDS

The two examples of § 605, one of the integers reduced modulo 4 and the other of the integers reduced modulo 5, display a structure bearing at least some superficial resemblance to the real number system. For example, there are elements apparently behaving like the real numbers zero and unity. It is the purpose of this section to investigate the extent to which the integers modulo m are similar to the real numbers, and the extent to which they contrast. One observation that we may make at the outset is that since every system of integers modulo m is finite and every ordered field is infinite (cf. Theorem V, § 312), *for no natural number m is the system of integers modulo m an ordered field.* Our next question is: "Under what circumstances, if any, is the system of integers modulo m a field?" We shall find that the answer depends on the value of m.

It has already been observed in § 604, and is evident from the multiplication table of Example 1, § 605, that it is possible for the product of two integers to be congruent to zero modulo m even though neither factor by itself is congruent to zero modulo m. This shows that the system of integers modulo m is not always a field—for example, when $m = 6$ and when $m = 4$ it is not. What is there about the numbers 6 and 4 that makes the resulting system *not* a field?

Before attempting to answer the questions just raised, let us see what properties are true for every system of integers modulo m. We start by defining one of the most important concepts of abstract algebra.

Definition. *A **ring** is a set with two binary operations of addition and multiplication subject to Axioms I(i), (ii), (iii), (iv), and (v), II(i) and (ii), and III, § 102. A **commutative ring with unity** is a ring in which Axioms II(iii) and (v), § 102, hold: that is, it is subject to all of the Axioms I, II, and III of a field except for Axiom II(iv), which guarantees multiplicative inverses for nonzero elements.*†

NOTE 1. Every field is an integral domain, and every integral domain is a commutative ring with unity.

Theorem I. *If m is a natural number greater than 1, the system of integers modulo m is a commutative ring with unity.*

Proof. We take the axioms in turn. I(i): True by (i), § 605. I(ii): By the Definition, § 605, $[a] + ([b] + [c]) = [a] + [b + c] = [a + (b + c)] = [(a + b) + c] = [a + b] + [c] = ([a] + [b]) + [c]$. I($iii$): The element that plays the role of zero is the equivalence class $[0]$: $[a] + [0] = [a + 0] = [a]$. I($iv$): For a given equivalence class $[a]$ the equivalence class $[-a]$ plays the

---

† For an elementary treatment of the theory of rings the reader is referred to Neal H. McCoy, *Introduction to Modern Algebra* (Boston, Allyn and Bacon, Inc., 1960) and G. Birkhoff and S. MacLane, *A Survey of Modern Algebra* (New York, The Macmillan Co., 1953).

role of the additive inverse: $[a] + [-a] = [a - a] = [0]$. $\text{I}(v)$: $[a] + [b] =$ $[a + b] = [b + a] = [b] + [a]$. $\text{II}(i)$: True by (iii), § 605. $\text{II}(ii)$: True by techniques similar to those used for $\text{I}(ii)$. $\text{II}(iii)$: The element that plays the role of unity is the equivalence class $[1]$, which is different from $[0]$ since $m > 1$: $[a][1] = [a \cdot 1] = [a]$. $\text{II}(v)$: $[a][b] = [ab] = [ba] = [b][a]$. III: $[a]([b] + [c]) = [a][b + c] = [a(b + c)] = [ab + ac] = [ab] + [ac] = [a][b] +$ $[a][c]$.

We now are ready to provide answers to some questions asked earlier.

Theorem II. *If m is a natural number greater than 1, the ring of integers modulo m is a field if and only if m is prime.*

Proof. First assume that m is composite: $m = rs$, where r and s are natural numbers greater than 1. Then $[r][s] = [m] = [0]$ whereas, since $0 < r < m$ and $0 < s < m$, $[r] \neq [0]$ and $[s] \neq [0]$. But, by Example 4, § 103, in any field the product of two nonzero elements is nonzero. Therefore, the integers modulo m cannot be a field. Now assume that m is prime: $m = p$, and let $[a]$ be any nonzero equivalence class, where $0 < a < p$. Since the GCD of a and p is unity: $(a, p) = 1$, by Theorem V, § 408, there exist natural numbers r and s such that $ar + p(-s) = 1$. Therefore $[a][r] + [p][-s] =$ $[a][r] + [0][-s] = [a][r] = [1]$. The equivalence class $[r]$ is therefore the multiplicative inverse of $[a]$, and Axiom $\text{II}(v)$ is satisfied. Therefore the system of integers modulo p is a field.

NOTE 2. The number of members of a finite field is called its **order.** Theorem II tells us that for every prime p there is a finite field of order p. It is not true, however, that *every* finite field is of prime order. What is true is that every finite field is of prime-power order (that is, is of order p^n for some prime p and natural number n), and conversely, for every prime p and natural number n there is a finite field of order p^n. (For proofs, see the Birkhoff and MacLane book cited in the preceding footnote, pp. 445–447.) For example, there are no fields of order 6, 18, or 100, but there are fields of order 8, 9, and 625. The reader may be interested in verifying that the set consisting of the four distinct members 0, 1, a, and b, with the following addition and multiplication tables, is a field (cf. Ex. 29, § 609):

	0	1	a	b
0	0	1	a	b
1	1	0	b	a
a	a	b	0	1
b	b	a	1	0

Addition

	0	1	a	b
0	0	0	0	0
1	0	1	a	b
a	0	a	b	1
b	0	b	1	a

Multiplication

Example. Find the multiplicative inverse of 72 in the field $\mathscr{F}$ of integers modulo 89. Thereby find an integer n such that $0 < n < 89$ and $72n \equiv 13 \pmod{89}$.

Solution. The Euclidean Algorithm gives

$$89 = 72 + 17,$$
$$72 = 4 \cdot 17 + 4,$$
$$17 = 4 \cdot 4 + 1,$$

and hence $1 = 17 - 4(72 - 4 \cdot 17) = 17 \cdot 17 - 4 \cdot 72 = 17(89 - 72) - 4 \cdot 72 = 17 \cdot 89 - 21 \cdot 72 = 72(89 - 21) - 89(72 - 17) = 72 \cdot 68 - 89 \cdot 55.$ In other words, $72 \cdot 68 = 89 \cdot 55 + 1$, $72 \cdot 68 \equiv 1 \pmod{89}$, and 68 is the desired multiplicative inverse of 72. We solve the equation $72n = 13$ in the field $\mathscr{F}$: $n = 13 \cdot 72^{-1} = 13 \cdot 68 = 83$. Checking in the integral domain $\mathscr{I}$ of integers: If $n = 83$, then $72 \cdot 83 - 13 = 5963 = 89 \cdot 67$.

*607. IDEALS

In the general theory of rings a particular kind of substructure known as an *ideal* assumes prominence. Since, in this book, our interest in ideals will be confined to ideals in integral domains, we shall restrict the definition to this particular type of ring:

Definition. *Let I be an integral domain. A nonempty set A of I is an* **ideal** *if and only if it is closed* (i) *with respect to addition and* (ii) *with respect to multiplication by arbitrary members of I:* (i) *If $a \in A$ and if $b \in A$, then $a + b \in A$.* (ii) *If $a \in A$ and $x \in I$, then $ax \in A$.*

Example 1. If $I = \mathscr{I}$, the integral domain of integers, and if c is a fixed natural number, show that the set of all integers of the form nc, where n is an integer:

(1) $$\cdots, -3c, -2c, -c, 0, c, 2c, 3c, \cdots,$$

is an ideal.

Solution. The set (1) is closed with respect to addition, since $mc + nc = (m + n)c$. It is closed with respect to multiplication by any member x of I, since $(nc)x = (nx)c$.

Example 2. In any integral domain I, show that the set A whose only member is the zero element 0 of I is an ideal.

Solution. Trivial.

Theorem I. *Any ideal A in an integral domain I is closed with respect to subtraction; that is, if $a \in A$ and if $b \in A$, then $a - b \in A$.*

Proof. Since $b \in A$, and since $-1 \in I$, $b(-1) = -b \in A$, and therefore $a + (-b) = a - b \in A$.

A simple corollary of Theorem I is:

Theorem II. *Any ideal A in an integral domain I is a commutative additive group.* (Cf. Exs. 30, 31, § 104.)

Proof. Since A is closed with respect to addition, Axiom I(i), § 102, is satisfied. Since the associative law holds for I it holds for A, and Axiom I(ii), § 102, is satisfied. Since A is nonempty it contains a member a and hence, by Theorem I, the element $a - a = 0$ of I, and Axiom I(iii), § 102, is satisfied. If $a \in A$, then by Theorem I the difference $0 - a = -a \in A$ and Axiom I(iv), § 102, is satisfied. Finally, since I is commutative so is A, and Axiom I(v), § 102, is satisfied.

For ideals in the integral domain $\mathscr{I}$ of the integers, Examples 1 and 2 are representative:

Theorem III. *If $\mathscr{I}$ is the integral domain of the integers (in any ordered field), and if A is any ideal in $\mathscr{I}$ containing a nonzero member, then there exists a unique natural number c such that A is the ideal of Example 1, consisting of all integral multiples of c.*

Proof. Since A contains a nonzero member s, it must contain at least one positive member (either s or $-s$). Let c be the least positive member of A, and let B denote the ideal (1). Since $c \in A$ and since A is closed with respect to multiplication by arbitrary members of $\mathscr{I}$, $nc \in A$ for every integer n and we have proved that $B \subset A$. We shall now prove that $A \subset B$ by assuming that there is a member t of A that is not a member of B, and obtaining a contradiction. Since t is not a member of B, $t \neq 0$. If $t > 0$ let $a \equiv t$, and if $t < 0$ let $a \equiv -t$. In either case $a \in A$, $a > 0$, and a is not a member of B. By the Fundamental Theorem of Euclid (Theorem II, § 403) there exist nonnegative integers q and r such that $r < c$ and $a = qc + r$. Since $qc \in B$ and a is not a member of B, $r \neq 0$, and hence $0 < r < c$. On the other hand, since $a \in A$ and $qc \in A$, their difference is a member of A (Theorem I): $a - qc = r \in A$. The desired contradiction has been reached, since r is a positive member of A less than the least positive member c. This establishes the inclusion $A \subset B$ which, with $B \subset A$, gives the desired result: $A = B$. The uniqueness of c follows from the fact that there do not exist distinct natural numbers each of which is an integral multiple of the other.

The first part of the proof of Theorem III shows that the ideal (1) of Example 1 is the smallest ideal containing the number c in the sense that *any* ideal that contains c contains the ideal (1). The ideal (1) is said to be **generated** by the number c. Theorem III states that any ideal in $\mathscr{I}$ that contains more than the number 0 is generated by a unique natural number.

In a similar fashion, as the reader may readily verify, if a and b are any two natural numbers there exists a smallest ideal A in $\mathscr{I}$ containing a and b in the sense that any ideal containing a and b contains A. This ideal A, defined in the following example, is said to be **generated** by a and b:

Example 3. If $I = \mathscr{I}$, the integral domain of integers, and if a and b are any two fixed natural numbers, show that the set A of all integers of the form $ma + nb$, where m and n are integers, is an ideal.

Solution. The set A is closed with respect to addition, since $(m_1a + n_1b) + (m_2a + n_2b) = (m_1 + m_2)a + (n_1 + n_2)b$. It is closed with respect to multiplication by any member x of I, since $(ma + nb)x = (mx)a + (nx)b$.

We now come to one of the most interesting aspects of this whole discussion of ideals, the relation between generated ideals and the GCD:

Theorem IV. *If A is the ideal generated by the natural numbers a and b, and if c is the (single) natural number that generates A, then c is the GCD of a and b: $c = (a, b)$.*

Proof. Since A consists of all integral multiples of c and since both a and b are members of A, each is an integral multiple of c; in other words, c is a common divisor of a and b. On the other hand, since c is a member of the ideal A of Example 3, c has the form $c = ma + nb$, where m and n are integers, and any common divisor of a and b is a divisor of c. Therefore c is the *greatest* common divisor of a and b.

*608. LINEAR EQUATIONS

It was seen in Exercise 29, § 104, that in any field the general linear equation $ax + b = 0$, $a \neq 0$, has a unique solution $x = -b/a$. The finite fields of the integers modulo a prime number provide examples for illustrating this principle.

Example 1. Solve the equation

(1) $$3x + 4 = 0$$

in the system of integers modulo 5 (where brackets are omitted for simplicity).

Solution. From the multiplication table of Example 2, § 605, we see that 2 is the reciprocal of 3 since their product is 1. Therefore $x = -4/3 = -4 \cdot 2 = -3$. Since the additive inverse of 3 is 2, the solution of (1) is $x = 2$, as may be verified by direct substitution in (1).

Systems of equations can be handled in a finite field of integers modulo a prime much as they can in the field of real numbers. A full treatment would demand a study of determinants and matrices. The interested reader will find a discussion of these topics, and further references, in the two books cited in the footnote of § 606. A **solution** of a system of equations in the *variables* or *unknowns* $x, y, \cdots$ is a set of values of these variables that satisfy all equations simultaneously.

Example 2. Solve the system of equations

(2) $$\begin{cases} 2x + 4y = 3, \\ x + 3y = 2, \end{cases}$$

for x and y in the field of integers modulo 5.

Solution. The variable y can be eliminated by first multiplying the second equation of (2) on both sides by 2 and adding the resulting two members of the equation $2x + y = 4$ to those of the first equation, with the result: $4x = 2$, or $x = 3$. Substitution in the first equation gives $1 + 4y = 3$, $4y = 2$, $2y = 1$, and $y = 3$. The solution of (2) is therefore $x = 3$, $y = 3$, as can be verified by direct substitution.

Example 3. Show that the following system of equations:

(3)
$$\begin{cases} 3x + 2y + 3z = 1, \\ x + 3y + 2z = 3, \\ 2x + 4y + z = 4, \end{cases}$$

has no solution (x, y, z) in the field of integers modulo 5.

Solution. Addition of the first two equations gives $4x = 4$, or $x = 1$. Substitution gives for the first and third equations: $2y + 3z = 3$ and $4y + z = 2$, respectively. Doubling both members of the first of these two equations gives $4y + z = 1$, an equation incompatible with $4y + z = 2$.

Example 4. Show that the equation

$$6x = 8$$

has no solution in the ring of integers modulo 12. (Cf. Ex. 28, § 609.)

Solution. If x is an integer and if $(6x - 8)/12$ is an integer, then $(12x - 16)/12 = x - \frac{4}{3}$ is an integer. (Contradiction.)

Example 5. Find all solutions of the equation

(4)
$$6x = 8$$

in the ring of integers modulo 14. (Cf. Ex. 28, 609.)

Solution. We first seek *a* solution, and then *all* solutions. Since the GCD of 6 and 14 is 2, it is possible, by means of the Euclidean Algorithm (cf. § 408), to find natural numbers m and n such that $6m - 14n = 2$; in fact, with $m = 5$ and $n = 2$, $6m - 14n = 30 - 28 = 2$. Multiplying both members of $6 \cdot 5 - 14 \cdot 2 = 2$ by 4 gives $6 \cdot 20 - 14 \cdot 8 = 8$, or $6(14 + 6) - 14 \cdot 8 = 8$, and hence $6 \cdot 6 \equiv 8 \pmod{14}$, with a solution of $x = 6$ for the given equation. We now make use of the solution $x = 6$ to discover what other solutions $6 + a$ there may be, by studying the equation $6(6 + a) \equiv 8 \pmod{14}$, or $6a \equiv 0 \pmod{14}$. The integer a satisfies this last congruence if and only if $6a/14$, or $3a/7$ is an integer, and this is true if and only if a is a multiple of 7. If a is a multiple of 7 and if $6 + a$ is between 0 and 13, inclusive, then $a = 0$ or 7, and conversely. Therefore there are exactly two solutions of (4): $x = 6$ and $x = 13$.

*609. EXERCISES

In Exercises 1–4, solve the given congruence of the form $ax \equiv b \pmod{c}$ for x between 0 and c.

1. $8x \equiv 6 \pmod{11}$.   **2.** $1001x \equiv 1 \pmod{101}$.

3. $5x \equiv 3 \pmod{6}$.   **4.** $49x \equiv 20 \pmod{60}$.

In Exercises 5–8, find the multiplicative inverse of the given member a of the field of integers modulo the prime number p.

5. $a = 12, p = 13.$ **6.** $a = 21, p = 23.$

7. $a = 22, p = 37.$ **8.** $a = 12, p = 97.$

9. Find the integer n such that $0 < n < 11$ and $7^{37} \equiv n \pmod{11}$.

10. Find the integer n such that $0 < n < 19$ and $13^{37} \equiv n \pmod{19}$.

In Exercises 11–12, solve the simultaneous system in the field of integers modulo 7.

11.
$$\begin{cases} 5x + 3y = 6, \\ 6x + 2y = 5. \end{cases}$$

12.
$$\begin{cases} 2x - 6y + 5z = 3, \\ x + 3y - 5z = 3, \\ 5x + 2y + 3z = 2. \end{cases}$$

In Exercises 13–14, show that the given system of equations has no solution in the field of integers modulo 13.

13.
$$\begin{cases} 5x + 11y = 2, \\ 2x + 7y = 8. \end{cases}$$

14.
$$\begin{cases} x + y + z = 1, \\ x + 2y + 3z = 1, \\ 10x + 4y + 11z = 1. \end{cases}$$

In Exercises 15–16, find all solutions of the given system in the field of integers modulo 5.

15.
$$\begin{cases} 3x + 2y = 4, \\ 4x + y = 2. \end{cases}$$

16.
$$\begin{cases} x + y + 2z = 2, \\ x + 4y + 2z = 3, \\ 2x + y + 4z = 2. \end{cases}$$

17. Find all solutions of the equation $x^2 = 5$ in the field of integers modulo 11.

18. Find all solutions of the equation $x^2 = 5$ in the field of integers modulo 13.

19. Prove that if p is a prime greater than 3, then there exists an integer a between 1 and p such that the equation $x^2 = a$ has two solutions in the field of integers modulo p, and there exists an integer b between 1 and p such that the equation $x^2 = b$ has no solutions in the field of integers modulo p. (Cf. Exs. 17–18.)

20. Show that the ring of integers modulo m, where m is a natural number greater than 1, is an integral domain (cf. the Note, § 501) if and only if it is a field.

21. Discuss the relationship between the solutions of the equation $2x + 5y = 3$ in the ring of integers modulo 6 and the solutions of the equation $4x + 10y = 6$ in the ring of integers modulo 12.   State and prove a general principle illustrated by this example.

22. Give a reasonable definition for the product nx, where x is an arbitrary member of a field $\mathscr{F}$ and n is an arbitrary integer (n is not necessarily a member of $\mathscr{F}$).

Prove the following laws, where m and n denote arbitrary integers and x and y denote arbitrary members of $\mathscr{F}$: (i) $0 \cdot x = 0$, where the first 0 is the integer 0 and the second 0 is the zero of $\mathscr{F}$; (ii) $n \cdot 0 = 0$, where each 0 is the zero of $\mathscr{F}$; (iii) $1 \cdot x = x$; (iv) $(-1)x = -x$; (v) $m(nx) = (mn)x$; (vi) $m(xy) = (mx)y$; (vii) $(m + n)x = mx + nx$; (viii) $n(x + y) = nx + ny$.

23. Show that with proper interpretation the formula (10), § 314, for the sum of an arithmetic sequence remains valid in any field where $2 \equiv 1 + 1 \neq 0$. Illustrate by computing the following sum in the field of integers modulo 11 in two ways: $2 + 7 + 12 + 17 + \cdots + 102$. (Cf. Ex. 22.)

24. Show that with proper interpretation the formula (12), § 314, for the sum of a geometric sequence remains valid in any field. Illustrate by computing the following sum in the field of integers modulo 5 in two ways: $1 + 3 + 3^2 + 3^3 + \cdots + 3^{10}$. (Cf. Ex. 22.)

25. Let a and b be members of the set $\mathscr{N}$ of natural numbers, and let $a\rho b$ mean that either $a = b = 1$ or $a > 1$ and $b > 1$ and the greatest prime divisor of a is equal to the greatest prime divisor of b. (For example, $15 \rho 40$ since 5 is the greatest prime divisor of each of 15 and 40.) Prove that ρ is an equivalence relation. If the equivalence class of a is denoted $[a]$, consider the following attempt by Simple Simon to define an operation of addition for the equivalence classes: $[a] + [b] = [a + b]$. Prove that this operation is *not* well defined; that is, show that the equivalence class resulting from this combination of two equivalence classes may depend on the representatives chosen from them.

In Exercises 26–30, R denotes the ring of integers modulo m, where m is a natural number greater than 1.

26. Prove that a has a multiplicative inverse in R if and only if a and m are relatively prime.

27. If a and b are natural numbers between 1 and m, and if either a and m are relatively prime or b and m are relatively prime, prove that in R the equation $ax + by = c$ has exactly m solutions. Find all solutions of the equation $5x + 6y = 8$ in the ring of integers modulo 12.

28. Prove that the equation $ax = c$ has a solution in R if and only if $(a, m) \mid c$.

29. Verify that the system defined by the addition and multiplication tables of Note 2, § 606, is a field.

30. A member a of the ring R is a **null divisor** if and only if $a \neq 0$ and there exists a member b of R such that $b \neq 0$ and $ab = 0$. Prove that if a is a nonzero member of R, then a is a null divisor if and only if $(a, m) > 1$.

31. Discuss the concept of GCD for three natural numbers a, b, and c with the aid of the theory of ideals. Show that the GCD of a, b, and c can be expressed in the form $ma + nb + pc$, where m, n, and p are integers. Extend your discussion to the GCD of an arbitrary finite number of natural numbers. (Cf. Exs. 25–28, § 409.)

32. If a and b are integers and p is prime, prove that

$$(a + b)^p \equiv a^p + b^p \pmod{p}.$$

(Cf. Ex. 27, § 318.)

33. Prove **Fermat's Theorem** (due to the French mathematician Pierre Fermat (1601–1665)): *If a is a natural number and p is a prime number that does not divide a, then $a^{p-1} \equiv 1$* (mod *p*). Check Fermat's Theorem for a few values of *a* and *p*. *Hint:* Let $a_k \equiv ka$, $k = 1, 2, \cdots, p - 1$. Show that no two of these a_k's are congruent modulo *p*, so that they belong to the *p* − 1 equivalence classes of the natural numbers $1, 2, \cdots, p - 1$, and consequently

$$a_1 a_2 \cdots a_{p-1} = (p - 1)! \, a^{p-1} \equiv (p - 1)! \; (\text{mod } p).$$

7

*Polynomials and Rational Functions

++

*701. INTRODUCTION

If, in the real number system $\mathscr{R}$ (or, more generally, in any ordered field $\mathscr{G}$), we think of starting with the number 1 and generate by means of addition, subtraction, and multiplication the smallest system of numbers closed with respect to these three operations we obtain the integral domain $\mathscr{I}$ of the integers. If we include the fourth "rational operation," of division, we have the ordered field of rational numbers.

If, similarly, in the family of real-valued functions from $\mathscr{R}$ to $\mathscr{R}$ we begin with members of $\mathscr{R}$ and the special function f defined by the equation $f(x) = x$, and then generate by means of addition, subtraction, and multiplication the smallest system of functions closed with respect to these three operations we obtain a very important new integral domain. It is our first purpose in this chapter to study the members of this integral domain, known as *polynomials*. We shall see that they form a structure similar in many ways to that of $\mathscr{I}$, the set of integers. We shall then turn our attention to the *rational functions*, whose role among functions is analogous to that of the rational numbers among the real numbers.

*702. POLYNOMIALS

Definition I. *A polynomial†, or polynomial function, in the variable x is a function f on $\mathscr{R}$ into $\mathscr{R}$ defined by an expression of the form*

$$(1) \qquad f(x) = a_0 x^0 + a_1 x^1 + a_2 x^2 + \cdots + a_n x^n$$
$$= a_0 + a_1 x + a_2 x^2 + \cdots + a_n x^n,$$

† For a discussion of the algebraic concept of a polynomial as an abstract form, where the letter x is an "indeterminate" symbol rather than a member of $\mathscr{R}$ or any other particular set, see the references in the footnote of § 606. In the present text the single word *polynomial* is restricted to apply only to a polynomial function on $\mathscr{R}$ into $\mathscr{R}$. The statements of this chapter are valid if $\mathscr{G}$ is any ordered field, and practically all of them that do not involve the order relation of $\mathscr{G}$ are true if $\mathscr{G}$ is replaced by any *infinite* field. However, if the coefficients are drawn from a *finite* field the approach to the study of polynomials must be by indeterminants and not by functions.

where $x \in \mathcal{R}$ *and* $a_k \in \mathcal{R}$, $k = 0, 1, \cdots, n$. *The expression* (1) *is also called a* **representation** *of the polynomial* f. *The numbers* a_k, $k = 0, 1, \cdots, n$ *are called* **coefficients** *of* f, *in the representation* (1). *The number* a_k *is also called the* **coefficient** *of* x^k *and of the* **term** $a_k x^k$, $k = 0, 1, \cdots, n$. *The* **zero polynomial**, *denoted* 0 *and defined*

(2) $0(x) = 0$ *for every* $x \in \mathcal{R}$,

is defined by (1) *in case every coefficient vanishes:* $a_k = 0, k = 0, 1, \cdots, n$. *A* **nonzero polynomial** *is any polynomial other than the zero polynomial.*

Examples 1. The following are polynomials:

(i) $2 + 0x + 0x^2 + \frac{1}{2}x^3$,   (ii) $3x - \frac{2}{3}x^2 + 0x^3 + 0x^4$,
(iii) $-5 + \frac{1}{4}x^3 - 17x^{10}$,   (iv) 4.

Before we can proceed much further, even with definitions, it is necessary to establish an important basic uniqueness theorem (Theorem II, below). We start with a special case of that theorem. (Cf. § 705 for stronger forms of these theorems.)

Theorem I. *The only representations of the zero polynomial are those with every coefficient equal to zero. That is, if* h *is a polynomial defined:*

(3) $h(x) = c_0 + c_1 x + \cdots + c_r x^r$,

and if $h(x) = 0$ *for every* $x \in \mathcal{R}$, *then* $c_k = 0$ *for* $k = 0, 1, \cdots, r$.

Proof. Assume that there exists a representation (3) for the zero polynomial for which at least one coefficient is nonzero. Let n be the greatest subscript of the nonzero coefficients in (3), so that $h(x)$ can be written in the form $c_0 + c_1 x + \cdots + c_n x^n$, where $c_n \neq 0$. If $c_n > 0$, let $f \equiv h$, and if $c_n < 0$ let $f \equiv -h$. In either case, we have a polynomial of the form

(4) $f(x) = a_0 + a_1 x + \cdots + a_n x^n$,

where $a_n > 0$, and such that $f(x) = 0$ for all real numbers x. We shall obtain a contradiction to this as a consequence of the following lemma:

Lemma. *If* f *is a polynomial of the form* (4), *where* $a_n > 0$ *and* $n > 0$, *there exists a positive number* b *such that* $f(x) > 1$ *whenever* $x > b$.

Proof of Lemma. Let $P(n)$ be the proposition that the statement of the lemma is true for the positive integer n. Then $P(1)$ is true since the polynomial $a_0 + a_1 x$, where $a_1 > 0$, is greater than 1 in case $x > (1 - a_0)/a_1$. Now assume that $P(n)$ is true for a particular positive integer n, and consider $P(n + 1)$. Write the polynomial $f(x) = a_0 + a_1 x + \cdots + a_{n+1} x^{n+1}$, where $a_{n+1} > 0$, in the form:

(5) $f(x) = a_0 + x(a_1 + a_2 x + \cdots + a_{n+1} x^n)$.

By the induction assumption there exists a positive number b such that the quantity in parentheses in (5) is greater than 1 whenever $x > b$. If $x > \max(b, 1 - a_0)$, then $f(x) > a_0 + x(1) > a_0 + (1 - a_0) = 1$, and $P(n + 1)$ is true. By the Fundamental Theorem of Mathematical Induction, the proof of the lemma is complete.

Finally, since the special case $n = 0$ is trivial, and for $n > 0$ the statement of the lemma is stronger than needed, the proof of Theorem I is complete.

Theorem II. *Any two representations of a polynomial are identical except possibly for zero coefficients. That is, if f is a polynomial and if*

(6)
$$\begin{cases} f(x) = a_0 + a_1 x + a_2 x^2 + \cdots + a_m x^m, \\ f(x) = b_0 + b_1 x + b_2 x^2 + \cdots + b_n x^n, \end{cases}$$

where $m \leq n$, for every $x \in \mathcal{R}$, then $a_k = b_k$ for $k = 0, 1, \cdots, m$ and $b_k = 0$ for $k = m + 1, m + 2, \cdots, n$.

Proof. Letting $c_k \equiv b_k - a_k$ for $k = 0, 1, \cdots, m$ and $c_k \equiv b_k$ for $k = m + 1, m + 2, \cdots, n$, and forming the difference between the two right-hand members of (6), we obtain the following representation for the zero polynomial:

(7)
$$0(x) = c_0 + c_1 x + c_2 x^2 + \cdots + c_n x^n$$

By Theorem I, $c_k = 0$ for $k = 0, 1, \cdots, n$, and the proof is complete.

The import of Theorem II is that two polynomials are equal *as functions* if and only if they are represented by expressions (1) of the *same form*—except possibly for terms having zero coefficients. Thus, $x + 0x^2 - x^3$ and $x - x^3 + 0x^4$ are equal polynomials, whereas $x + x^2$ and $x + x^3$ are not equal polynomials. Theorem II makes the following definition possible without ambiguity:

Definition II. *Let f be a nonzero polynomial in the variable x, and let $f(x)$ be expressed in the form (1). The **degree** of f, written $\deg f$ or $\deg f(x)$, is the exponent of the highest power of x, appearing in (1), that has a nonzero coefficient. The zero polynomial has no degree. A polynomial of degree zero is a **nonzero constant polynomial**. A polynomial of positive degree is a **nonconstant polynomial**. A polynomial of degree 1, 2, 3, 4, or 5 is a **linear, quadratic, cubic, quartic,** or **quintic** polynomial, respectively. The coefficient of x^n in a polynomial of degree n is called the **leading coefficient**. The coefficient of x^0—that is, in (1), the number a_0—is called the **constant term**.*

It was observed in the Note, § 501, that the integers constitute an integral domain; that is, they form a system that satisfies all of the axioms of a field enumerated in § 102 except that the existence of a reciprocal is replaced by the weaker assumption that the product of any two nonzero numbers is nonzero. Using the operations of addition and multiplication of functions given in Note 3, § 310, we shall presently prove (Theorem III, below) a similar result for polynomials.

Examples 2. If f and g are the polynomials

$$f(x) = 1 + 3x - x^2, \qquad g(x) = 2 + x^2 - 4x^3,$$

then $f + g$ and fg are the polynomials

$$f(x) + g(x) = 3 + 3x - 4x^3, \quad f(x)g(x) = 2 + 6x - x^2 - x^3 - 13x^4 + 4x^5.$$

Theorem III. *If f and g are any two polynomials on $\mathscr{R}$ into $\mathscr{R}$ and if their sum $p = f + g$ and product $q = fg$ are defined:*

(8) $$p(x) = f(x) + g(x), \qquad q(x) = f(x)g(x),$$

then the set of all polynomials on $\mathscr{R}$ into $\mathscr{R}$ is an integral domain. The degrees of $f + g$ and fg are subject to the following laws whenever the quantities involved exist (that is, whenever the zero polynomial is not present):

(9) $$\deg (f + g) \leq \max (\deg f, \deg g), \qquad \deg (fg) = \deg f + \deg g.$$

Proof. Showing that the sum and product of two polynomials are polynomials is a matter of repeated use of associative, commutative, and distributive laws of real numbers with suitable collection of terms, as illustrated by Examples 2. The details are left to the reader. Checking the axioms defining an integral domain is almost completely routine. For example, the commutative, associative, and distributive laws hold since they hold for the real number system in which the polynomial functions have their values (these three laws, in fact, hold for any real-valued functions defined on any common domain). The zero of Axiom I(*iii*) is the zero polynomial. The negative of $a_0 + a_1x + \cdots + a_nx^n$ is $-a_0 - a_1x - \cdots - a_nx^n$. The unity of Axiom II(*iii*) is the nonzero constant polynomial 1. If f and g are nonzero polynomials with leading coefficients a_m and b_n, respectively, then their product fg has the nonzero leading coefficient a_mb_n and is therefore a nonzero polynomial. This last fact shows that whenever $\deg f$ and $\deg g$ exist, so does $\deg (fg)$, and $\deg (fg) = m + n = \deg f + \deg g$. The inequality in (9) follows from the obvious impossibility of the contrary inequality $\deg (f + g) > \max (\deg f, \deg g)$.

Corollary. *In the integral domain of polynomials the cancellation law for multiplication holds: If f is not the zero polynomial, the equality $fg = fh$ implies the equality $g = h$.*

Proof. This follows from Theorem III by the Note, § 501.

*703. DIVISORS AND MULTIPLES

Nearly all of Chapter 4, which concerns problems of divisibility among natural numbers, has a companion theory for polynomials. In this section and its sequel we shall sketch some of this theory. When the analogy is almost complete, we shall avoid a dull repetition of discussion already given and leave such details to the reader. It might be well at this point for the reader to look once more at the first paragraph of § 604 concerning divisibility among integers.

Definition I.   *A nonzero polynomial f **divides** a polynomial h (in symbols: $f \mid h$) if and only if there exists a polynomial g such that $h = fg$. If $f \mid h$, f is is called a **divisor** or **factor** of h, and h is called a **multiple** of f. A polynomial h is called **reducible** or **composite** if and only if there exist nonconstant polynomials f and g such that $h = fg$. A polynomial f is **irreducible** or **prime** if and only if f is of positive degree and not reducible. Two nonzero polynomials are **relatively prime** if and only if their only common divisors are constants.*

Many elementary properties of polynomials are similar to those listed in Theorem I, § 401, and in the first paragraph of § 604. In particular, every nonzero polynomial f is a divisor of the zero polynomial 0: $f \mid 0$. If c is any nonzero constant polynomial and if f is any nonzero polynomial, then $c \mid f$ and $f \mid f$. If $f \mid g$ ana $g \mid f$, then each of f and g is a nonzero constant times the other, and conversely. If f and g are nonzero polynomials of equal degree and if $f \mid g$, then each of f and g is a nonzero constant times the other, and conversely. If f is the polynomial $f(x) = x$, then $f \mid g$ if and only if the constant term of g is 0. Every linear polynomial is prime. If $f \mid g$ then $\deg f \leq \deg g$.

Example 1.   Prove that every reducible polynomial has degree at least 2. Show that $x^2 + 1$ is prime.

Solution. If h is reducible and $h = fg$, where f and g are nonconstant polynomials, then $\deg h = \deg f + \deg g \geq 1 + 1 = 2$. If $x^2 + 1$ were reducible it would have the form $x^2 + 1 = (ax + b)(cx + d) = acx^2 + (ad + bc)x + bd$. By Therem II, § 702, $ac = bd = 1$ and $ad + bc = 0$. Therefore $bc = -ad$ and $(ac)(bd) = (ad)(bc) = -(ad)^2 = 1$, or $(ad)^2 = -1$, in contradiction to the fact that the square of any nonzero number is positive (Example 5, § 203).

Theorem I.   *Every nonconstant polynomial has a nonconstant divisor of least degree. This divisor is prime. Therefore every nonconstant polynomial has at least one prime divisor.*

Proof. By analogy with the proof of Theorem II, § 401, we let A be the set of all natural numbers that are degrees of divisors of the given nonconstant polynomial f. Since f divides itself and has positive degree, A is nonempty and contains a least member m. If g is any divisor of f having degree m, then g must be prime, since otherwise we could write $g = \phi\psi$, where the degrees of ϕ and ψ would be natural numbers whose sum is m so that each would be a nonconstant polynomial divisor of f of degree less than the least possible!

Theorem V, § 401, becomes (with almost identical proof):

Theorem II.   *Every nonconstant polynomial is a product of a finite number of prime polynomials.*

Proof.   Let f be a polynomial of positive degree n, and let A be the set of all natural numbers m such that m is the degree of a divisor g of f that is a product of a finite number of prime polynomials.   By Theorem I, A is nonempty.   Since $1 \leqq m \leqq n$, the set A is finite and has a greatest member k. We wish to show that $k = n$.   Assume $k < n$.   Then there exists a divisor g of f that has degree k and is a product of prime polynomials.   Since f has the form $f = gh$ and $\deg f = k + \deg h$, h has positive degree and hence (Theorem I) a prime divisor ϕ.   Consequently $g\phi$ is a divisor of f that is a product of prime factors and has degree greater than k, which was assumed to be the largest possible such degree.   Finally, with $k = n$ established, we have $f = cg$, where c is a nonzero constant and g is a product of prime factors.   Letting c be incorporated as a part of one of these factors, we have f equal to a product of prime polynomials, as desired.

*704. THE FUNDAMENTAL THEOREM OF EUCLID. THE DIVISION ALGORITHM

For polynomials the Fundamental Theorem of Euclid is formulated as follows:

Theorem.   Fundamental Theorem of Euclid.   *If f is any polynomial and g is any nonzero polynomial, then there exist unique polynomials q and r such that*

(1) $$r = 0 \quad or \quad \deg r < \deg g$$

and

(2) $$f(x) = q(x)g(x) + r(x), \qquad x \in \mathcal{R}.$$

Proof.   We start by proving uniqueness.   Let $f(x) = q_1(x)g(x) + r_1(x) = q_2(x)g(x) + r_2(x)$.   Then $r_1(x) - r_2(x) = (q_2(x) - q_1(x))g(x)$.   If $q_2 - q_1$ were not the zero polynomial we should have $\deg(r_1 - r_2) = \deg(q_2 - q_1) + \deg g$, and hence $\deg(r_1 - r_2) \geqq \deg g$, an inequality inharmonious with (1). Consequently $q_2 - q_1 = 0$, and hence $r_1 - r_2 = 0$.   That is, $q_1 = q_2$ and $r_1 = r_2$, and uniqueness is proved.   We now turn our attention to existence. There are two cases.   If g divides f, then f has the form $f = qg$, or (2) with $r = 0$, and there is no more to prove.   Assume, therefore, that g is *not* a divisor of f.   Let A be the set of all nonnegative integers that are degrees of polynomials of the form $f - qg$, where q is an arbitrary polynomial.   Then A is nonempty (for example, the polynomial $f - 1 \cdot g$ has a degree since, with g not a divisor of f, $f - g$ is not the zero polynomial), and hence has a least member m.   We shall show that $m < \deg g$ by assuming $m \geqq n \equiv \deg g$ and obtaining contradiction.   Accordingly, let q be any polynomial such that $r = f - qg$ is a polynomial of degree m, with leading coefficient a_m, and let the leading coefficient of g be b_n. With the assumption that $m \geq n$ we can subtract from both sides of the equation $f(x) - q(x)g(x) = r(x)$ the expression $(-a_m/b_n)x^{m-n}g(x)$ to obtain

(3) $\qquad f(x) - \left(q(x) + \dfrac{a_m}{b_n} x^{m-n}\right) g(x) = r(x) - \dfrac{a_m}{b_n} x^{m-n}g(x).$

Since g is not a divisor of f the right-hand member of (3) is not the zero polynomial. On the other hand, its degree is less than that of r, as is seen by noting that the highest degree term of $r(x)$ has been removed by subtraction. Thereby a contradiction has been obtained, since m was assumed to be the *least* nonnegative integer that is the degree of any expression of the form (3). Finally, if r denotes a polynomial of least degree having the form $f - qg$, then the conditions (1) and (2) of the theorem are satisfied, and the proof is complete.

NOTE. The polynomial r of the preceding theorem is called the **remainder** when f is divided by g. The polynomial q is called the **quotient**, and f and g are called the **dividend** and the **divisor**, respectively. As in the case of numbers, the word *quotient* is also used in the sense of the ratio of two functions, f/g, defined by $f(x)/g(x)$. In any individual case the context should indicate which usage is intended. In this book (unless a contrary statement is made) *the single word quotient should be interpreted to mean a ratio.*

In practice the quotient q and remainder r, of formula (2), are obtained by a standard process of long division. This process is called the *division algorithm*, and is illustrated in the following examples.

Example 1. Find the quotient q and the remainder r of formula (2) if $f(x) = 6x^4 + x^2 - 7$ and $g(x) = 2x^2 + 8x - 5$.

Solution. The first step of the standard division process appears thus:

(4)
$$
\begin{array}{r}
3x^2 \\
2x^2 + 8x - 5\overline{)6x^4 + x^2 - 7} \\
6x^4 + 24x^3 - 15x^2 \\
\hline
-24x^3 + 16x^2 - 7
\end{array}
$$

and states that

(5) $\qquad 6x^4 + x^2 - 7 = 3x^2(2x^2 + 8x - 5) + (-24x^3 + 16x^2 - 7).$

This has the form of (2), but condition (1) is not satisfied, since $\deg(-24x^3 + 16x^2 - 7) > \deg(2x^2 + 8x - 5)$. The next step in the division is

(6)
$$
\begin{array}{r}
-12x \\
2x^2 + 8x - 5\overline{)-24x^3 + 16x^2 - 7} \\
-24x^3 - 96x^2 + 60x \\
\hline
112x^2 - 60x - 7
\end{array}
$$

whence $(-24x^3 + 16x^2 - 7) = -12x(2x^2 + 8x - 5) + (112x^2 - 60x - 7)$.   In conjunction with (5) this gives:

(7) $6x^4 + x^2 - 7 = (3x^2 - 12x)(2x^2 + 8x - 5) + (112x^2 - 60x - 7)$.

Since deg $(112x^2 - 60x - 7) = $ deg $(2x^2 + 8x - 5)$, condition (1) is still not satisfied and we continue once more:

$$
\begin{array}{r}
56 \\
2x^2 + 8x - 5 \overline{)112x^2 - 60x - 7} \\
112x^2 + 448x - 280 \\
\hline
- 508x + 273
\end{array}
$$

(8)

and (5) takes the form

(9) $6x^4 + x^2 - 7 = (3x^2 - 12x + 56)(2x^2 + 8x - 5) + (-508x + 273)$.

Therefore $q(x) = 3x^2 - 12 + 56$ and $r(x) = -508x + 273$.   Notice that deg $r = 1 < \deg g = 2$.

In practice, the three division steps are combined into a single familiar algorithm:

$$
\begin{array}{r}
3x^2 - 12x + 56 \\
2x^2 + 8x - 5 \overline{)6x^4 + + x^2 -7} \\
6x^4 + 24x^3 - 15x^2 \\
\hline
-24x^3 + 16x^2 \\
-24x^3 - 96x^2 + 60x \\
\hline
112x^2 - 60x \\
112x^2 + 448x - 280 \\
\hline
-508x + 273.
\end{array}
$$

Example 2.   Find the quotient q and the remainder r of formula (2) if $f(x) = 2x^3 + x^2 - 7x - 11$ and $g(x) = 3x^2 - x - 2$.

Solution.   It is clear at the outset that the first coefficient of q involves division by 3 and that there are two steps to the division algorithm.   To avoid fractions we multiply f by 9, and suppress the powers of x in the following form:

$$
\begin{array}{r}
6 \quad 5 \\
3 \quad -1 \quad -2 \overline{)18 \quad 9 \quad -63 \quad -99} \\
18 \quad -6 \quad -12 \\
\hline
15 \quad -51 \\
15 \quad -5 \quad -10 \\
\hline
-46 \quad -89.
\end{array}
$$

This states that

$$18x^3 + 9x^2 - 63x - 99 = (6x + 5)(3x^2 - x - 2) + (-46x - 89),$$

or, if both members are divided by 9:

$$2x^3 + x^2 - 7x - 11 = (\tfrac{2}{3}x + \tfrac{5}{9})(3x^2 - x - 2) + (-\tfrac{46}{9}x - \tfrac{89}{9}).$$

Therefore $q(x) = \tfrac{2}{3}x + \tfrac{5}{9}$ and $r(x) = -\tfrac{46}{9}x - \tfrac{89}{9}$.

*705. THE REMAINDER AND FACTOR THEOREMS

The Fundamental Theorem of Euclid takes a particularly simple but useful form in case the divisor is linear with leading coefficient 1, and it gives us:

Theorem I. Remainder Theorem. *If a polynomial $f(x)$ is divided by a linear polynomial of the form $x - c$, where c is a real number, until a constant remainder R is obtained, this remainder is equal to the value of the function f when x is equal to c:*

$$(1) \qquad\qquad R = f(c).$$

Proof. By (1), § 704, the remainder in the Theorem, § 704, is a constant $R = r(x)$, and by (2), § 704,

$$(2) \qquad\qquad f(x) = q(x)(x - c) + R,$$

for every real number x. In particular, (2) holds for the value $x = c$: $f(c) = q(c)(c - c) + R$, and (1) results.

As a corollary to Theorem I we have the well-known *factor theorem*. For the sake of simplicity of statement we define at this time a **root** or **zero** of a polynomial $f(x)$ to be a real number c such that $f(c) = 0$.

Theorem II. Factor Theorem. *If $f(x)$ is a polynomial and c is a real number, then the linear polynomial $x - c$ is a divisor (or factor) of $f(x)$ if and only if c is a root (or zero) of $f(x)$: $f(c) = 0$.*

Proof. The polynomial $x - c$ is a divisor of $f(x)$ if and only if the remainder R in (2) is equal to 0. But since, by Theorem I, $R = f(c)$, R vanishes if and only if $f(c)$ vanishes: $f(c) = 0$, and the proof is complete.

NOTE. The factor theorem provides us with a new means of proving that any two representatives of a polynomial must have identical nonzero coefficients (Theorem II, § 702). This is a corollary of the following theorem:

Theorem III. *A polynomial of degree n cannot have more than n distinct roots.*

Proof. Let $P(n)$ be the proposition that the statement of the theorem is correct for the nonnegative integer n. For $n = 0$ the proposition is trivial since it states that a nonzero constant is never zero. For $n = 1$ the proposition merely states that a linear equation $ax + b = 0$ $(a \neq 0)$ cannot have two distinct roots; this fact is a consequence of Exercise 29, § 104. Now assume that $P(n)$ is true for a particular natural number n, and consider a polynomial of degree $n + 1$:

$$(3) \qquad f(x) = a_0 + a_1 x + \cdots + a_n x^n + a_{n+1} x^{n+1},$$

where $a_{n+1} \neq 0$. Assume now that $f(x)$ has m distinct roots $c_1, c_2, \cdots, c_m$, where $m > n + 1$. In particular, the real number c_m is a root and hence, by the factor theorem, $f(x)$ has the quantity $(x - c_m)$ as a divisor and can thus be written:

(4) $f(x) = (x - c_m)g(x)$,

where $g(x)$ is a polynomial of degree n. By the induction assumption, $g(x)$ has at most n roots. But this contradicts our assumption that $c_1, c_2, \cdots,$ c_{m-1}, where $m - 1 > n$, are all roots of $g(x)$ since, by (4), $f(c_k) = (c_k - c_m)g(c_k) = 0$ implies $g(c_k) = 0$ since $c_k - c_m \neq 0$, for $k = 1, 2, \cdots,$ $m - 1$. With this contradiction, and use of the Fundamental Theorem of Mathematical Induction, the proof is complete.

Corollary. *If f and g are nonzero polynomials, each of degree at most n, and if $f(x) = g(x)$ for more than n distinct values of x, then $f(x) = g(x)$ for all real numbers x, and $f = g$.*

Proof. If $f(x) = g(x)$ for more than n distinct values of x and $f \neq g$, then the nonzero polynomial $f - g$, of degree at most n, has more than n distinct roots.

Example 1. If $f(x) = 5x^3 - 8x - 1$, find $f(-2)$ by division.

Solution. By the Remainder Theorem we find the remainder after $f(x)$ is divided by $x - (-2) = x + 2$:

$$
\begin{array}{r}
5x^2 - 10x + 12 \\
x + 2)\overline{5x^3 \qquad\quad - 8x - 1} \\
5x^3 + 10x^2 \\
\hline
-10x^2 \\
-10x^2 - 20x \\
\hline
12x \\
12x + 24 \\
\hline
-25.
\end{array}
$$

Therefore $f(-2) = -25$.

Example 2. Show that $x - 1$ is a factor of $f(x) = x^{100} - 37x^{19} + 36$.

Solution. By the Factor Theorem, since $f(1) = 0$, 1 is a root of f and hence $x - 1$ is a factor.

*706. DIVISIBILITY OF PRODUCTS

We adapt for the integral domain of polynomials on $\mathscr{R}$ into $\mathscr{R}$ most of the theorems and some of the proofs of § 404.

Theorem I. *A prime polynomial p cannot be a divisor of the product of two polynomials f and g each of which is a nonzero polynomial of degree less than that of p.*

Proof. Assume that $p \mid fg$ and that $\deg f < \deg p$ and $\deg g < \deg p$. Let h be a nonzero polynomial of *least* degree such that $p \mid fh$. Then

$0 < \deg h \leqq \deg g < \deg p$. (If the degree of h were 0, then h would be a nonzero constant and the degrees of p and f would be equal.) Let q and r be polynomials, according to the Fundamental Theorem of Euclid (§ 704), such that $p = qh + r$, where $\deg r < \deg h$. (Since p is prime, r cannot be the zero polynomial.) Multiplying by the polynomial f and rearranging terms, we have $fr = fp - q(fh)$, which is divisible by p. In other words, r is a polynomial of degree less than that of h such that $p \mid fr$, whereas h was assumed to be a polynomial of the *least* possible degree such that $p \mid fh$. With this contradiction the proof is complete.

Theorem II. *If p is a prime polynomial, and if f and g are polynomials such that $p \mid fg$, then $p \mid f$ or $p \mid g$ (or both).*

Proof. Assume that p is a divisor of fg but a divisor of *neither f* nor g. Then both f and g are nonzero polynomials and, by the Fundamental Theorem of Euclid, there exist polynomials q, r, s, and t, where r and t are nonzero polynomials of degree less than that of p, such that $f = qp + r$ and $g = sp + t$. Then, since $fg = (qsp + qt + rs)p + rt$ it follows that $p \mid rt$, in contradiction to Theorem I.

As a corollary (see Theorem III, § 404, for a proof) we have:

Theorem III. *A prime polynomial that divides the product of n polynomials must divide at least one of the n factors.*

It is left to the reader to verify that the details of the proof of Theorem IV, § 404, can be adapted—almost without change—to a proof of the following theorem:

Theorem IV. *If f, g, and h are polynomials, if h and f are relatively prime, and if $h \mid fg$, then $h \mid g$.*

*707. UNIQUE FACTORIZATION THEOREM

As might be expected, the Unique Factorization Theorem for natural numbers has its counterpart for polynomials. Both in the statement of the following theorem and in its proof the inequalities of § 405 are replaced by inequalities among the *degrees* of the polynomials involved. The details are left for the reader. The Unique Factorization Theorem for polynomials is a uniqueness theorem for the representation of a nonconstant polynomial as a product of prime polynomial factors, guaranteed to exist by Theorem II, § 703.

Theorem. Unique Factorization Theorem. *The representation of a nonconstant polynomial as a product of prime polynomials is unique except for the order of the factors and the multiplication by nonzero constants. If the prime factors are arranged according to the order of their degrees the representation is unique except for nonzero constant multipliers. In other words, if*

$p_1, p_2, \cdots, p_r$ and $q_1, q_2, \cdots, q_s$ are prime polynomials of (positive) degrees $m_1 \leq m_2 \leq \cdots \leq m_r$ and $n_1 \leq n_2 \leq \cdots \leq n_s$, respectively, and if

$$(1) \qquad p_1 p_2 \cdots p_r = q_1 q_2 \cdots q_s,$$

then $r = s$ and there exist nonzero constants $c_1, c_2, \cdots, c_r$ such that $p_k = c_k q_k$ for $1 \leq k \leq r = s$.

★708. THE EUCLIDEAN ALGORITHM

The Euclidean Algorithm for polynomials (cf. § 407) takes the following form (whose derivation is left to the reader—Ex. 7, § 710):

Theorem. Euclidean Algorithm. *If f and g are nonzero polynomials, with $\deg f \geq \deg g$, and if g does not divide f, then there exist nonzero polynomials $q_1, q_2, \cdots, q_{k+1}$ and $r_1, r_2, \cdots, r_k$, where $k \geq 1$, such that*

$$(1) \qquad \left\{ \begin{array}{ll} f = q_1 g + r_1, & \deg r_1 < \deg g, \\ g = q_2 r_1 + r_2, & \deg r_2 < \deg r_1, \\ r_1 = q_3 r_2 + r_3, & \deg r_3 < \deg r_2, \\ \qquad \cdots & \\ r_{k-2} = q_k r_{k-1} + r_k, & \deg r_k < \deg r_{k-1}, \\ r_{k-1} = q_{k+1} r_k. & \end{array} \right.$$

In case $k = 1$, the last equation of (1) should be interpreted as $r_0 \equiv g = q_2 r_1$. In case $g \mid f$, system (1) reduces to the single equation $f = q_1 g$ which, with $r_{-1} \equiv f$ and $r_0 \equiv g$, is the last equation of (1) with $k = 0$.

Example. Apply the Euclidean Algorithm to the polynomials $3x^3 + 5x^2 + 5x + 2$ and $12x^2 - 7x - 10$.

Solution. We start by dividing the cubic polynomial by the quadratic:

$$\begin{array}{r} \frac{1}{4}x + \frac{9}{16} \\ 12x^2 - 7x - 10 \overline{)3x^3 - 5x^2 + 5x + 2} \\ \underline{3x^3 - \frac{7}{4}x^2 - \frac{5}{2}x} \\ \frac{27}{4}x^2 + \frac{15}{2}x \\ \underline{\frac{27}{4}x^2 - \frac{63}{16}x - \frac{45}{8}} \\ \frac{183}{16}x + \frac{61}{8}. \end{array}$$

The remainder is $\frac{61}{16}(3x + 2)$. To simplify the next step we temporarily set aside the constant factor $\frac{61}{16}$, and divide $12x^2 - 7x - 10$ by $3x + 2$:

$$\begin{array}{r} 4x - 5 \\ 3x + 2 \overline{)12x^2 - 7x - 10} \\ \underline{12x^2 + 8x} \\ -15x \\ \underline{-15x - 10} \\ 0. \end{array}$$

Putting the two divisions together, we have for the Euclidean Algorithm in the form (1), with $f(x) = 3x^3 + 5x^2 + 5x + 2$ and $g(x) = 12x^2 - 7x - 10$:

$$3x^3 + 5x^2 + 5x + 2 = (\tfrac{1}{4}x + \tfrac{9}{61})(12x^2 - 7x - 10) + \tfrac{61}{16}(3x + 2),$$
$$12x^2 - 7x - 10 = [\tfrac{16}{61}(4x - 5)][\tfrac{61}{16}(3x + 2)].$$

*709. GCD AND LCM

The discussion of GCD and LCM for natural numbers (§ 408) has its parallel for polynomials. The principal difference is that with polynomials greatest common divisors and least common multiples are not unique without special restriction. To this end we define a **monic polynomial** to be a polynomial whose leading coefficient is equal to 1.

In this section the proofs are omitted since the details are mere adaptations of those of § 408. These details may be supplied by the reader. (Exs. 8–12, § 710.)

Definition I. *A nonzero polynomial h is called* **a greatest common divisor**, *or a* GCD *for short, of the two nonzero polynomials f and g if and only if it has the following two properties:*

(i) $h \mid f$ *and* $h \mid g$;

(ii) *whenever* ϕ *is a nonzero polynomial and* $\phi \mid f$ *and* $\phi \mid g$, *then* $\phi \mid h$.

The greatest common divisor, *or the* GCD *for short, of f and g is the monic polynomial H that is a* GCD *of f and g, with the notation*

(1)   *The* GCD *of f and* $g = H = (f, g)$.

Theorem I. *If f and g are any two nonzero polynomials, the GCD of f and g exists and is unique. Any two greatest common divisors of f and g are nonzero constant multiples of each other. In the notation of the Euclidean Algorithm (Theorem, § 708), if c is the leading coefficient of* r_k,

$$(f, g) = \frac{1}{c} r_k.$$

Example 1. By the Example, § 708, the GCD of $3x^3 + 5x^2 + 5x + 2$ and $12x^2 - 7x - 10$ is $\tfrac{1}{3}(3x + 2) = x + \tfrac{2}{3}$.

Theorem II. *Two nonzero polynomials f and g are relatively prime if and only if the GCD of f and g is 1:* $(f, g) = 1$.

Theorem III. *If f and g are nonzero polynomials, if h is a GCD of f and g, and if* $f = f_1 h$ *and* $g = g_1 h$, *then* f_1 *and* g_1 *are relatively prime:* $(f_1, g_1) = 1$.

Definition II. *A nonzero polynomial v is called* **a least common multiple**, *or an* LCM *for short, of the nonzero polynomials f and g if and only if it has the following two properties:*

(i) $f \mid v$ *and* $g \mid v$;

(ii) *whenever* ϕ *is a nonzero polynomial and* $f \mid \phi$ *and* $g \mid \phi$, *then* $v \mid \phi$.

The least common multiple, or the LCM *for short, of f and g is the monic polynomial V, that is an* LCM *of f and g, with the notation*

(2)   The LCM *of f and g* $= V = [f, g]$.

Theorem IV. *If f and g are any two nonzero polynomials, the* LCM *of f and g exists and is unique. Any two least common multiples of f and g are nonzero constant multiples of each other. The* GCD *and* LCM *of f and g are related:*

(3) $$(f, g) [f, g] = \frac{1}{c} fg,$$

where c is the leading coefficient of fg.

Example 2. By Example 1 and Theorem IV,

$$[3x^3 + 5x^2 + 5x + 2, 12x^2 - 7x - 10]$$

$$= \frac{1}{12} \frac{(3x^3 + 5x^2 + 5x + 2)(12x^2 - 7x - 10)}{3x + 2}$$

$$= \frac{1}{12} (12x^4 + 5x^3 - 5x^2 - 17x - 10).$$

Continuing the analogy with § 408, we have the representation of any GCD of two nonzero polynomials f and g as a linear combination of f and g:

Theorem V. *If f and g are nonzero polynomials and if h is a* GCD *of f and g, then there exist polynomials ϕ and ψ such that*

(4) $$\phi f + \psi g = h.$$

The nonzero polynomials f and g are relatively prime if and only if there exist polynomials ϕ and ψ such that

(5) $$\phi f + \psi g = 1.$$

NOTE. The polynomials ϕ and ψ of Theorem V are by no means unique since, if ϕ and ψ satisfy (4) or (5), so do $\phi + \chi g$ and $\psi - \chi f$ for an arbitrary polynomial χ.

Example 3. Find polynomials ϕ and ψ such that

$$\phi(x)(3x^3 + 5x^2 + 5x + 2) + \psi(x)(12x^2 - 7x - 10) = 3x + 2.$$

Solution. From the first part of the solution of the Example, § 708, we have

$$3x^3 + 5x^2 + 5x + 2 = (\tfrac{1}{4}x + \tfrac{9}{16})(12x^2 - 7x - 10) + \tfrac{61}{16}(3x + 2).$$

Therefore

$$3x + 2 = \tfrac{16}{61}(3x^3 + 5x^2 + 5x + 2) + (-\tfrac{4}{61}x - \tfrac{9}{61})(12x^2 - 7x - 10),$$

and a determination of ϕ and ψ is given by

$$\phi(x) = \tfrac{16}{61}, \qquad \psi(x) = -\tfrac{4}{61}x - \tfrac{9}{61}.$$

Example 4. Find polynomials ϕ and ψ such that
$$\phi(x)(x^3 + 1) + \psi(x)(x^4 + 1) = 1.$$

Solution. The Euclidean Algorithm gives
$$\begin{cases} x^4 + 1 = x(x^3 + 1) + (-x + 1), \\ x^3 + 1 = (-x^2 - x - 1)(-x + 1) + 2, \\ -x + 1 = (-\tfrac{1}{2}x + \tfrac{1}{2})2, \end{cases}$$
from which we get:
$$2 = (x^3 + 1) + (x^2 + x + 1)[(x^4 + 1) - x(x^3 + 1)]$$
$$= (-x^3 - x^2 - x + 1)(x^3 + 1) + (x^2 + x + 1)(x^4 + 1).$$
We can therefore choose:
$$\phi(x) = -\tfrac{1}{2}x^3 - \tfrac{1}{2}x^2 - \tfrac{1}{2}x + \tfrac{1}{2}, \qquad \psi(x) = \tfrac{1}{2}x^2 + \tfrac{1}{2}x + \tfrac{1}{2}.$$

*710. EXERCISES

1. Give an example of nonzero polynomials f and g such that $f + g$ is nonzero and deg $(f + g) <$ deg f and deg $(f + g) <$ deg g.

2. Prove that if f is a prime polynomial and if c is a nonzero constant then cf is a prime polynomial.

3. Prove that $x^2 + x + 1$ is prime.

In Exercises 4–5, find the quotient q and the remainder r of formula (2), § 704, for the given polynomials f and g.

4. $f(x) = 3x^2 + 2x + 1, g(x) = 4x^2 + 3x + 2$.

5. $f(x) = x^4 - x^3 + x^2 - x + 1, g(x) = 2x^2 - 1$.

6. Use the factor theorem to prove the special case of Theorem II, § 706, where p is a linear monic polynomial.

7. Prove the Theorem, § 708.　　　　**8.** Prove Theorem I, § 709.

9. Prove Theorem II, § 709.　　　　**10.** Prove Theorem III, § 709.

11. Prove Theorem IV, § 709.　　　　**12.** Prove Theorem V, § 709.

In Exercises 13–14, find a GCD and an LCM of the two given polynomials.

13. $x^4 + x^3 + 2x^2 - x + 3, 3x^4 - x^3 + 2x^2 + x + 1$.

14. $x^5 - 3x^2 - 25x - 15, x^6 - 5x^4 - 15x - 9$.

In Exercises 15–16, find polynomials ϕ and ψ such that $\phi f + \psi g = 1$, for the given polynomials f and g of the indicated Exercises.

15. Exercise 4.　　　　　　**16.** Exercise 5.

17. If f, g, and h are nonzero polynomials, if f and h are relatively prime, and if g and h are relatively prime, prove that fg and h are relatively prime. If f and h are relatively prime and n is a natural number, prove that f^n and h are relatively prime.

18. If f, g, and h are nonzero polynomials, if f and g are relatively prime, if $f \mid h$, and $g \mid h$, prove that $fg \mid h$.

19. Prove that polynomials ϕ and ψ exist such that $\phi f + \psi g = h$, where f, g, and h are given nonzero polynomials, if and only if $(f, g) \mid h$.

20. Prove the rational roots theorem of college algebra: *If f is a polynomial*:

(1) $$f(x) = a_0 + a_1x + a_2x^2 + \cdots + a_nx^n,$$

all of whose coefficients $a_0, a_1, a_2, \cdots, a_n$ are integers, if $a_0 \neq 0$, if $a_n \neq 0$, and if $r = p/q$ is a rational root of f ($f(r) = 0$), where p and q are integers and the fraction p/q is in lowest terms, then $p \mid a_0$ and $q \mid a_n$. Hint: Write $q^n f(p/q) = 0$ in the forms:

$$q(a_0 q^{n-1} + \cdots + a_{n-1} p^{n-1}) = -a_n p^n,$$
$$p(a_1 q^{n-1} + \cdots + a_n p^{n-1}) = -a_0 q^n,$$

and use the ideas of Exercise 15, § 409, and Exercise 17 above.

21. As a corollary of Exercise 20, prove that there is no rational number whose square is equal to 2. (Cf. Theorem III, § 905.)   Prove that there is no rational number whose square is any one of the following: 3, 5, 6, 7, 8, 10, 11, 12.   (Cf. Exs. 5–6, § 1010.)

22. If I is the integral domain of polynomials and if A is any ideal in I containing a nonzero member, prove that there exists a unique monic polynomial h such that A consists of all polynomials of the form ϕh, where ϕ is an arbitrary polynomial.   (Cf. Theorem III, § 607.)

23. In the integral domain I of polynomials, discuss the concepts of the ideal generated by a nonzero polynomial, the ideal generated by two nonzero polynomials, and state and prove the analogue of Theorem IV, § 607.

24. Adapt Exercise 31, § 609, to the integral domain of polynomials.

*711. RATIONAL FUNCTIONS

If f and g are any two nonzero polynomials, then the function ϕ defined by the equation

$$\text{(1)} \qquad\qquad \phi(x) = \frac{f(x)}{g(x)}$$

for all numbers x for which the denominator is nonzero, has a domain of definition that consists of all but a finite number† of real numbers (Theorem III, § 705).   If f and g have a nonconstant common divisor h, with $f = f_1 h$ and $g = g_1 h$, then a reduction of the fraction in (1) to lower terms by cancellation of the common factor $h(x)$ (whenever $h(x) \neq 0$) leads to a simplified equation

$$\text{(2)} \qquad\qquad \psi(x) = \frac{f_1(x)}{g_1(x)},$$

where the polynomials f_1 and g_1 are relatively prime (Theorem III, § 709). The values of the function ψ of (2) are equal to those of ϕ for all x in the domain of ϕ, since $h(x) \neq 0$ whenever $g(x) \neq 0$.   However, the *function* ψ might not be the same as the *function* ϕ since the domain of ψ might be larger than that of ϕ.

Example 1. If $\phi(x) = \dfrac{x}{x(x-1)}$, and if $\psi(x) = \dfrac{1}{x-1}$, then the domain of ϕ consists of all real numbers with the two exceptions 0 and 1.   The domain of ψ, on the other hand, includes the number 0 as well.

---

† The empty set is considered to be a finite set with the finite number of 0 elements.   The entire set $\mathscr{R}$ of real numbers, then, consists of all but a finite number of real numbers.

Since, in defining a function, it is necessary to have the domain of definition —as well as the values—specified, it is important to ensure uniqueness of domain for any function defined by an equation of the form (1). Our main objective in this section will be to study and resolve this problem. We start with some definitions involving ordered pairs of polynomials. The notation $f:g$ is suggested by (1), and is used to avoid confusion with the notation (f, g) for the GCD of f and g.

Definition I. *Whenever f and g are polynomials, the expression $f:g$ denotes the ordered pair whose first coordinate is f and whose second coordinate is g, where g is a nonzero polynomial. The ordered pair $f:g$ is **in lowest terms** if and only if either $f = 0$ and g is a nonzero constant or f and g are relatively prime nonzero polynomials. The ordered pair $f:g$ is **in canonical form** if and only if $f:g$ is in lowest terms and g is monic.*

Example 2. In Example 1, $x:x(x - 1)$ is an ordered pair of polynomials that is not in lowest terms; $1:(x - 1)$ is in lowest terms and in canonical form. The following are in lowest terms but not in canonical form:

$$1:(2x - 1), \quad 0:3, \quad (5x - 1):(1 - x).$$

Definition II. *Equivalence between ordered pairs of polynomials is defined as follows: $f:g$ is **equivalent** to $r:s$, written*

(3) $$f:g \sim r:s,$$

if and only if $fs = gr$.

Example 3. The first two ordered pairs of Example 2, drawn from Example 1, are equivalent.

Theorem I. *The relation defined by (3) is an equivalence relation.*

Proof. It is obvious that $f:g \sim f:g$ and that if $f:g \sim r:s$ then $r:s \sim f:g$. To prove transitivity we assume that $f:g \sim r:s$ and that $r:s \sim u:v$; in other words, that $fs = gr$ and $rv = su$. After multiplication by v and g, respectively, these equations become $fsv = grv$ and $grv = gsu$. Consequently, $fsv = gsu$ and by the cancellation law for multiplication, since $s \neq 0$, $fv = gu$ and $f:g \sim u:v$.

A connection between equivalence of ordered pairs of polynomials and proportionality of their values is given in the theorem:

Theorem II. *If f and r are any polynomials and g and s are any nonzero polynomials, then*

(i) $$f:g \sim r:s$$

if and only if

(ii) $$\frac{f(x)}{g(x)} = \frac{r(x)}{s(x)} \text{ for infinitely many } x.$$

Proof. If (*i*) is true, then $f(x)s(x) = g(x)r(x)$ for all real numbers and hence for all numbers x for which $g(x)s(x) \neq 0$; hence for all but a finite number of real numbers x; hence for infinitely many x. Division by the nonzero quantity $g(x)s(x)$ gives (*ii*). On the other hand, if (*ii*) holds, then the polynomials fs and gr have equal values for infinitely many x: $f(x)s(x) = g(x)r(x)$ and hence, by the Corollary to Theorem III, § 705, must be identical: $fs = gr$, so that $f:g \sim r:s$.

Theorem III. *Ordered pairs of polynomials have the following properties involving equivalence:*

(*i*) *If* $h \neq 0$, $f:g \sim fh:gh$.

(*ii*) *If f and g are polynomials and if $g \neq 0$, there exist polynomials F and G, where $G \neq 0$, such that $f:g \sim F:G$ and $F:G$ is in lowest terms.*

(*iii*) *If f and g are polynomials and if $g \neq 0$, there exist unique polynomials F and G, where $G \neq 0$, such that $f:g \sim F:G$ and $F:G$ is in canonical form.*

Proof. (*i*): $fgh = fgh$. (*ii*): If $f = 0$, then $0:g \sim 0:1$. If $f \neq 0$, let $h = (f, g)$, $f = hF$, and $g = hG$. Then $(F, G) = 1$ and $f:g \sim hF:hG \sim F:G$, by (*i*). (*iii*): If $f = 0$, $0:g \sim 0:1$, and 1 is the only constant monic polynomial. If $f \neq 0$, let h in the proof of (*ii*) have leading coefficient equal to that of g. Then G is monic. Finally, if $F:G \sim R:S$, where $(F, G) = (R, S) = 1$ and G and S are monic, from $FS = GR$ we infer (by means of Theorem IV, § 706) that $G \mid S$ and $S \mid G$ and therefore that $G = S$; consequently $F = R$.

We return to the matter of defining rational functions.

Definition III.† *A* **rational function** *is a real-valued function ϕ of a real variable x whose value for a given x is given by an expression of the form*

$$(4) \qquad\qquad \phi(x) = \frac{F(x)}{G(x)},$$

where $F:G$ is in canonical form, and whose domain of definition is the set of all real numbers such that $G(x) \neq 0$.

It is clear from Definition III that any ordered pair of polynomials $F:G$ in canonical form uniquely determines a rational function ϕ by means of formula (4). The converse is also true.

Theorem IV. *If $F:G$ and $R:S$ are ordered pairs of polynomials in canonical form and if ϕ is a rational function such that $\phi(x) = F(x)/G(x)$ whenever $G(x) \neq 0$ and such that $\phi(x) = R(x)/S(x)$ whenever $S(x) \neq 0$, then $F:G = R:S$; that is, $F = R$ and $G = S$.*

† A definition equivalent to Definition III is obtained by replacing the words *canonical form* by the words *lowest terms*. The present formulation has been chosen in order to ensure uniqueness of the representation (4) (cf. Theorem IV).

Proof. By Theorem II, $F:G \sim R:S$ and hence, by part (*iii*) of Theorem III, $F = R$ and $G = S$.

Since any ordered pair $f:g$ of polynomials the second of which is nonzero uniquely determines an ordered pair $F:G$ which is equivalent to $f:g$ and is in canonical form, the ordered pair $f:g$ uniquely determines a rational function that is related to it in a natural way. We specify this correspondence, and introduce a special notation for it, in the following definition:

Definition IV. *If f is any polynomial and if g is any nonzero polynomial, let $F:G$ be the (unique) ordered pair of polynomials equivalent to $f:g$ and in canonical form, and let ϕ be the rational function defined by (4). The symbol $[f:g]$ denotes this rational function:*

$$(5) \qquad\qquad [f:g] \equiv \phi.$$

A few elementary properties of the correspondence that determines by means of (5) a rational function ϕ in terms of a given ordered pair $f:g$ of polynomials ($g \neq 0$) are given in the theorem:

Theorem V. *If f and r are any polynomials, if g, h, and s are any nonzero polynomials, and if 0 and 1 denote the constant polynomials identically equal to the real numbers 0 and 1, respectively, then:*

   (*i*) $[f:g] = [r:s]$ *if and only if* $f:g \sim r:s$;
   (*ii*) $[fh:gh] = [f:g]$;
   (*iii*) $[f:1] = f$;
   (*iv*) $[h:h] = [1:1] = 1$;
   (*v*) $[0:g] = [0:1] = 0$.

Proof. (*i*): If $[f:g] = [r:s] = \phi$, then, by Theorem IV, the ordered pairs $f:g$ and $r:s$ are separately equivalent to the same ordered pair in canonical form, and hence are equivalent to each other: $f:g \sim r:s$. On the other hand, if $f:g \sim r:s$, then the unique ordered pair $F:G$ in canonical form to which $f:g$ is equivalent must be equivalent to $r:s$ and hence must be the unique ordered pair $R:S$ in canonical form to which $r:s$ is equivalent. Therefore $[f:g] = [r:s]$. (*ii*): True by part (*i*), Theorem III. (*iii*): Since $f:1$ is in lowest terms and 1 is monic, if $\phi = [f:1]$, then for *every* real number x (the polynomial 1 never vanishes) the following equation holds: $\phi(x) = f(x)/1 = f(x)$, and $\phi = f$. (*iv*): By (*ii*) and (*iii*), $[h:h] = [1:1] = 1$. (*v*): By (*ii*) and (*iii*), $[0:g] = [0:1] = 0$.

Finally, a useful criterion for (5) to hold, expressed in terms of the values of the functions concerned, is given in the theorem:

Theorem VI. *The equation $\phi = [f:g]$, where ϕ is a rational function, holds if and only if the equation between real numbers:*

$$(6) \qquad\qquad \phi(x) = \frac{f(x)}{g(x)}$$

holds for infinitely many real numbers x.

Proof. Assume that a rational function ϕ is given in the form $\phi = [R:S]$, where $R:S$ is in canonical form, and that equation (6) holds for infinitely many values of x. Since $\phi(x) = R(x)/S(x)$ for all but a finite number of real numbers x (that is, except when $S(x) = 0$—cf. Theorem III, § 705), there must be infinitely many x such that the three quantities $\phi(x)$, $f(x)/g(x)$, and $R(x)/S(x)$ are all defined and equal. Therefore, by Theorem II, $f:g \sim R:S$ and hence, by part (i) of Theorem V, $[f:g] = [R:S] = \phi$. On the other hand, if $\phi = [f:g] = [F:G]$, where $F:G$ is in canonical form, then (by part (i), Theorem V), $f:g \sim F:G$, and $fG = gF$. Therefore the equation $f(x)G(x) = g(x)F(x)$ holds for all x and certainly for all x except for the finite set for which $g(x)G(x) \neq 0$. Thus, the equation $f(x)/g(x) = F(x)/G(x) = \phi(x)$ holds whenever the quantities involved are defined; hence, for all but a finite number of x; hence, for infinitely many x.

NOTE. For convenience, such a phrase as "the rational function $x^2/(x + 1)$" will be considered as an abbreviation for "the rational function $\phi = [f:g]$, where f and g are polynomials defined by the equations $f(x) = x^2$ and $g(x) = x + 1$."

*712. THE FIELD OF RATIONAL FUNCTIONS

In most cases, when functions are combined by addition or multiplication they are assumed to have a common domain of definition. For example, when polynomials are added and multiplied they have the entire real number system as their common domain, and the definitions and properties of addition and multiplication (Theorem III, § 702) are relatively simple.

With rational functions, however, the situation is quite different. There is no common domain of definition and, in fact, there can be no common domain of definition (unless by some irrelevant and inappropriate artifice such as assigning the value 0 whenever a function is not otherwise defined). For example, the rational function $\phi(x) = 1/(x - a)$ is undefined when x is equal to the real number a, and we conclude that *every* real number is excluded from the domain of definition of *some* rational function. On the other hand, we wish to show in the first part of this section that the rational functions form a field. This will mean, in particular, that every rational function other than the one that is identically zero must have a reciprocal. For example, the function $x/(x - 1)$ must have a reciprocal (presumably $(x - 1)/x$) such that the product (in some sense) is equal to the rational function 1, which is equal to the number 1 for *all* real numbers x, including both 0 and 1. (!) It was for the purpose of resolving such embarrassing paradoxes as this that such an elaborate treatment of rational functions was presented in § 711.

We start by defining the operations of addition and multiplication for rational functions, and then show that these definitions are meaningful (unambiguous). After this we shall show that the rational functions form a field, and finally, in § 713, that they form an ordered field. The ordered field of rational functions will be denoted $\mathscr{H}$.

Definition I. *Addition* and *multiplication* of rational functions $[f:g]$ and $[r:s]$ are defined as follows:

(i) $[f:g] + [r:s] \equiv [fs + gr:gs]$;

(ii) $[f:g] \cdot [r:s] \equiv [fr:gs]$.

Theorem I. *The definitions of addition and multiplication given in Definition I are independent of the particular polynomials f, g, r, and s used. That is, if $[p:q] = [f:g]$ and if $[u:v] = [r:s]$, then $[pv + qu:qv] = [fs + gr:gs]$, and $[pu:qv] = [fr:gs]$.*

Proof. By (i), Theorem V, § 711, let us assume that $gp = fq$ and $su = rv$, and seek to establish the two equalities $gs(pv + qu) = qv(fs + gr)$ and $(pu)(gs) = (qv)(fr)$. The first of these is obtained as follows: $gs(pv + qu) = (gp)(sv) + (gq)(su) = (fq)(sv) + (gq)(rv) = qv(fs + gr)$, while the second results thus: $(pu)(gs) = (gp)(su) = (fq)(rv) = (qv)(fr)$.

In the proof of the next theorem a lemma will be convenient:

Lemma. $[f:g] + [h:g] = [(f + h):g]$.

Proof. $[f:g] + [h:g] = [(fg + gh):g^2] = [g(f + h):g^2] = [(f + h):g]$, by (ii), Theorem V, § 711.

Theorem II. *The rational functions form a field. That is, with the definitions of addition and multiplication of Definition I, the Axioms I, II, and III of § 102 are satisfied. The additive identity, or zero, is the rational function $[0:1] = 0$, which is identically equal to 0 for all real numbers, and the additive inverse of $[f:g]$ is $[-f:g]$, so that we may write $-[f:g] = [-f:g]$. The rational function $[f:g]$ is equal to zero ($[f:g] = [0:1]$) if and only if $f = 0$; equivalently, $[f:g]$ is nonzero if and only if f is nonzero. The multiplicative identity, or unity, is the rational function $[1:1] = 1$. which is identically equal to 1 for all real numbers, and the multiplicative inverse of the nonzero $[f:g]$ is $[g:f]$, so that we may write $[f:g]^{-1} = [g:f]$.*

Proof. I(i): This was proved in Theorem I. I(ii): $[f:g] + ([p:q] + [r:s]) = [f:g] + [(ps + qr):qs] = [(f(qs) + g(ps + qr)):g(qs)] = [(fqs + gps + gqr):gqs] = [((fq + gp)s + (gq)r):(gq)s] = [(fq + gp):gq] + [r:s] = ([f:g] + [p:q]) + [r:s])$. I(iii): $[f:g] + [0:1] = [(f \cdot 1 + g \cdot 0):g \cdot 1] = [f:g]$. I(iv): $[f:g] + [(-f):g] = [(f - f):g] = [0:g] = [0:1]$, by the Lemma, and Theorem V, § 711. I(v): $[f:g] + [r:s] = [(fs + gr):gs] = [r:s] + [f:g]$. II(i): This was proved in Theorem I. II(ii): $[f:g]([p:q][r:s]) = [f:g][pr:qs] = [fpr:gqs] = [fp:gq][r:s] = ([f:g][p:q])[r:s]$. II(iii): $[1:1] \neq [0:1]$, since $1 \cdot 1 \neq 1 \cdot 0$, and $[f:g][1:1] = [f \cdot 1:g \cdot 1] = [f:g]$. II(iv): If $[f:g] \neq [0:1]$, then $f \cdot 1 \neq g \cdot 0$ and $f \neq 0$; also, $[f:g][g:f] = [fg:fg] = [1:1]$, by Theorem V, § 711. II(v): $[f:g][r:s] = [fr:gs] = [rf:sg] = [r:s][f:g]$. III: $[f:g]([p:q] + [r:s]) = [f:g][(ps + qr):qs] = [(fps + fqr):gqs] = [fps:gqs] +$

$[fqr:gqs] = [fp:gq] + [fr:gs] = [f:g][p:q] + [f:g][r:s]$ (with the aid of the Lemma).

It has already been indicated, in part (*iii*) of Theorem V, § 711, that the field of rational functions includes the integral domain of polynomials as a subset by means of the correspondence

(1) $$[f:1] \leftrightarrow f.$$

This is a one-to-one correspondence, since for all real numbers x the values of the rational function $[f:1]$ and the values of the polynomial f are identical. It is in this sense that we write equality:

(2) $$[f:1] = f.$$

In the spirit of algebraic structure, the correspondence (1) is much more than a mere correspondence since there are two operations, addition and multiplication, that apply both to rational functions and to polynomials. The basic question is this: "Suppose we have two functions each of which can be regarded as either a rational function or a polynomial, and suppose we wish to add (or multiply) these two functions. Does it matter whether we add (or multiply) them as rational functions or as polynomials?" Happily, the answer is in the negative:

Theorem III. *Let f and g be any polynomials, and let $[f:1]$ and $[g:1]$ be their corresponding rational functions, according to the correspondence* (1). *Then the sum and product of f and g as polynomials correspond to the sum and product of $[f:1]$ and $[g:1]$ as rational functions:*

(3) $$[f:1] + [f:1] \leftrightarrow f + g, \qquad [f:1][g:1] \leftrightarrow fg.$$

That is, the correspondence (1) *is an isomorphism with respect to addition and multiplication.* (cf. §§ 311, 502, 505, 906.)

Proof. By the Lemma to Theorem II, and by definition of multiplication of rational functions:

$$[f:1] + [g:1] = [(f + g):1], \qquad [f:1][g:1] = [fg:1].$$

The result just obtained can be expressed in three equivalent ways: (*i*) *In the field of rational functions those rational functions of the form $[f:1]$ form an integral domain* **isomorphic** *to the integral domain of the polynomials.* (*ii*) *The integral domain of the polynomials is* **embedded** *in the field of the rational functions.* (*iii*) *The field of the rational functions is an* **extension** *of the integral domain of the polynomials.*

NOTE. By virtue of Theorem III, it is immaterial, when we are adding or multiplying functions, whether we refer to the polynomials f and g as polynomials or as rational functions. The notations $[f:1]$ and f can be considered alternative methods of representing the rational function $[f:1]$.

We have just seen what happens when we add or multiply rational functions $[f:1]$ and $[g:1]$ that correspond to polynomials f and g. A similar statement holds for subtraction: $[f:1] - [g:1] \leftrightarrow f - g$. What happens with division? The following theorem shows that any rational function can be considered as a fraction whose numerator and denominator are polynomials (the denominator being nonzero). This is strictly analogous to the parallel situation in the real number system, where every rational number is equal to a fraction whose numerator and denominator are integers (the denominator being nonzero).

Theorem IV. *If $[f:g]$, $g \neq 0$, is an arbitrary rational function, then, in the field $\mathscr{H}$ of rational functions,*

$$(4) \qquad [f:g] = \frac{[f:1]}{[g:1]} = \frac{f}{g} .$$

Proof. Since, by Theorem II, $[g:1]^{-1} = [1:g]$, $[f:1]/[g:1] = [f:1][1:g] = [f:g]$.

*713. THE ORDERED FIELD OF RATIONAL FUNCTIONS

We have come now to the final part of our study of rational functions—the part having to do with *order*. As stepping stones toward proving that the field of rational functions is an ordered field we introduce (and justify) a definition, and establish a lemma.

Definition I. *The subset $\mathscr{P}$ of the field $\mathscr{H}$ of rational functions consists of all rational functions $[f:g]$ such that f and g are nonzero and have leading coefficients of the same sign.*

Theorem I. *Definition I is independent of the polynomials f and g used in its formulation. That is, if $f:g \sim r:s$, if $f \neq 0$, and if the leading coefficients of f and g have the same sign, then $r \neq 0$ and the leading coefficients of r and s have the same sign.*

Proof. If $f \neq 0$ and $fs = gr$, then $r \neq 0$. Assume that the leading coefficient a_m of f and the leading coefficient b_n of g have the same sign, and let c_j and d_k be the leading coefficients of r and s, respectively. From the equation $fs = gr$ we can infer in particular that the leading coefficient $a_m d_k$ of fs must be equal to the leading coefficient $b_n c_j$ of gr, and hence that $a_m/b_n = c_j/d_k$. Since a_m/b_n is assumed to be positive we conclude that c_j and d_k are of the same sign, as desired.

Lemma. *For any member $[f:g]$ of $\mathscr{P}$, f and g can be chosen so that both have positive leading coefficients.*

Proof. If $[f:g] \in \mathscr{P}$, and if both f and g have negative leading coefficients multiplying both f and g by (-1), and use the fact that $[-f:-g] = [f:g]$.

Theorem II.   *The field $\mathscr{H}$ of rational functions is an ordered field.*

Proof.   We must show that the set $\mathscr{P}$ of Definition I satisfies the three parts of Axiom IV, § 202.   For IV(i) and (ii), for any two given members of $\mathscr{P}$, $[f:g]$ and $[r:s]$, choose f, g, r, and s so that all four leading coefficients are positive.   Then the leading coefficients of the three polynomials $fs + gr$, fr, and gs are positive and consequently both $[f:g] + [r:s] = [(fs + gr):gs]$ and $[f:g][r:s] = [fr:gs]$ are members of $\mathscr{P}$.   Finally, for IV(iii), let $[f:g]$ be an arbitrary rational function, and assume (without loss of generality) that the leading coefficient of g is positive.   Then there are exactly three possibilities for f, exactly one of which is true: (first possibility) f is nonzero and its leading coefficient is positive, and $[f:g] \in \mathscr{P}$; (second possibility) $f = 0$, and $[f:g]$ is the zero rational function; (third possibility) f is nonzero and its leading coefficient is negative, and $-[f:g] = [-f:g] \in \mathscr{P}$.

NOTE 1.   From the preceding considerations we see that *the rational function $[f:g]$ is positive if and only if the leading coefficients of f and g have the same sign, and $[f:g]$ is negative if and only if the leading coefficients of f and g have opposite signs.*   In particular, *a nonzero polynomial is positive or negative according as its leading coefficient is positive or negative.*

As in Chapter 2, a transitive order relation satisfying the law of trichotomy is defined for the field of rational functions.

Example 1. $\dfrac{x^2 + 1}{3x - 1} < \dfrac{2x^4 - 1}{x^2 + 4}$, since the difference $\dfrac{2x^4 - 1}{x^2 + 4} - \dfrac{x^2 + 1}{3x - 1}$, when expressed in the form $\dfrac{6x^5 - 3x^4 - 5x^2 - 3x - 3}{(x^2 + 4)(3x - 1)}$, has positive leading coefficients in both numerator and denominator.

Example 2. $\dfrac{-x}{3x - 1} < \dfrac{2 - x}{3x}$, since the difference $\dfrac{2 - x}{3x} - \dfrac{-x}{3x - 1}$, when expressed in the form $\dfrac{7x - 2}{3x(3x - 1)}$, has positive leading coefficients in both numerator and denominator.

It will be shown later (§ 903), as a consequence of the Axiom of Completeness (§ 902) that the real number system has the property specified in the following definition:

Definition II.   *An ordered field $\mathscr{G}$ is **Archimedean** if and only if for any two positive members a and b of $\mathscr{G}$ there exists a natural number n such that $na > b$.   An ordered field is **non-Archimedean** if and only if it fails to be Archimedean.*

Theorem III.   *The ordered field $\mathscr{H}$ of rational functions is non-Archimedean.*

Proof. Let $a = 1$ and let b be the linear polynomial f where $f(x) = x$. Then for every natural number n, $na < b$ since the difference $b - na$ is a linear polynomial with positive leading coefficient 1.

Example 3. Let f and g be any two nonzero polynomials with positive leading coefficients such that $\deg f < \deg g$, and let c be an arbitrary positive real number. Prove that

(1)
$$0 < \frac{f}{g} < c < \frac{g}{f}.$$

Solution. The first inequality is true by definition. The second inequality is true since

$$\frac{c}{1} - \frac{f}{g} = \frac{cg - f}{g},$$

and the numerator and denominator both have positive leading coefficients. The third inequality follows similarly from

$$\frac{g}{f} - \frac{c}{1} = \frac{g - cf}{f}.$$

Note 2. Example 3 shows that in the ordered field of rational functions there is an infinite set located between zero and the set of positive real numbers, and there is another infinite set every member of which is greater than every real number.

For further discussion of non-Archimedean ordered fields the reader is referred to the following books and article, and the references contained therein: B. L. Van der Waerden, *Modern Algebra* (New York, Frederick Ungar Publishing Co., 1949), pp. 208–211, Paul Dubreil, *Algèbre* (Paris, Gauthiers-Villars, 1946), pp. 163–185, and B. H. Neumann, "On Ordered Division Rings," *Transactions of the American Mathematical Society*, volume 66 (1949), pp. 202–252.

*714. EXERCISES

1. Show that a positive rational function ϕ may have *some* negative values: $\phi(x) < 0$. Prove that if $\phi > 0$, then there exists a real number N such that $\phi(x) > 0$ whenever $x > N$.

2. If c is any real number, construct an infinite class A of rational functions ϕ such that for each member ϕ of A the following inequalities hold for every real number d greater than c:

$$c < \phi < d.$$

3. Simple Simon suggests the following operation on rational functions:

$$[f{:}g] \otimes [r{:}s] \equiv [f + r{:}gs].$$

Comment.

8

Intervals and Absolute Value

801. THE REAL AXIS

Geometrical intuition can be one of the most potent guides for mathematical study and discovery. At the same time it can be one of the most deceptive of counselors. Throughout the first seven chapters of this book we have assiduously avoided drawing figures or pictures in order to concentrate attention on the axioms and their consequences without subjecting ourselves to irrelevant and possibly improper influence. The time ultimately arrives, however, when a geometrical graph—kept under proper control—can be more of an asset than a liability. The reader should bear constantly in mind that whatever geometry appears in this book is incidental to the mathematical analysis, and should be regarded as a servant rather than master. These remarks are not intended to detract in any way from the status and dignity of geometry as a mathematical discipline, but rather, to point up the analytic and algebraic emphasis of the present volume. Although the *ideas* of some of our proofs may stem from geometry, their *proofs* must be shorn of their intuitive aspects and formulated in terms of the original axioms, or in terms of theorems derived deductively from those axioms. For discussion of the axiomatic foundations of geometry, the reader is referred to the following books: D. Hilbert, *The Foundations of Geometry* (La Salle, Ill., The Open Court Publishing Company, 1938), R. L. Wilder, *The Foundations of Mathematics* (New York, John Wiley & Sons, Inc., 1952), and H. Eves and C. V. Newsom, *An Introduction to the Foundations and Fundamental Concepts of Mathematics* (New York, Holt, Rinehart and Winston, 1953).

The real numbers are commonly represented by means of points on a line in a one-to-one correspondence with the real numbers. This line is said to be provided with a **number scale**, and is called the **real axis**. It is usually taken to be "horizontal"—and it is not our responsibility here to be concerned with the meaning of the word *horizontal*, or even to care whether it has any absolute mathematical meaning at all! The positive numbers correspond to points to the "right" of the point 0 that represents the number 0, and the negative numbers correspond to points to the "left" of 0. More generally, if x and y are any two *numbers* such that $x < y$ (or, equivalently, $y > x$), then the corresponding *points* x and y are located so that x is to the left of y, (or, equivalently, y is to the right of x). The points representing integers are spaced out so that the distance between 0 and 1 is the same as the distance between n and $n + 1$ for every integer n—where we have again represented points by the symbols normally attached to the corresponding points. More generally, in terms of the distance between 0 and 1 as a unit, the distance between any two points x and y, where $x < y$, is $y - x$. Figure 801 indicates the real axis, with a few points specifically labeled.

For convenience, much of the notation and terminology of real numbers will be used for points of the real axis. The context should dispel any

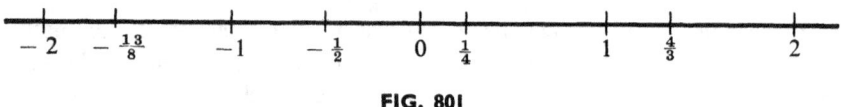

FIG. 801

possible misinterpretation. One instance is our use of the same symbol, say x, to refer to a number and to its corresponding point on the real axis as well. Another is the use of the word *between*: if the point x is to the left of the point z (that is, the numbers x and z bear the relation $x < z$) then the point y is between x and z (y is to the right of x and to the left of z) if and only if the number y is between the numbers x and z: $x < y < z$.

Further properties of the real number system, as deduced from the Axiom of Completeness in Chapter 9, will demonstrate the appropriateness of the real axis (Fig. 801) as a visual representation of the real numbers. Without the axiom of completeness the real axis may be a very improper model for an ordered field, as indicated in the following note:

NOTE. The ordered field $\mathcal{H}$ of rational functions (§ 713) cannot be represented on a line in the manner shown in Figure 801. For example, the linear polynomial x is greater than every natural number and would need to be represented by a point lying to the right of every point of Figure 801, and hence at an infinite distance. Also, the rational function $1/x$ satisfies the inequalities $0 < 1/x < 1/n$ for every positive integer n, and would therefore need to be represented by a point lying to the right of 0 and to the left of $1/n$ for every positive integer n, whereas between 0 and every point to the right of 0 lies a point of the form $1/n$. At the root of our paradoxical troubles is the fact that the system of rational functions is non-Archimedean.

802. FINITE INTERVALS

In this and the following section various types of sets, known as *intervals*, will be defined. Each of these intervals is defined by means of inequalities and is therefore formulated most naturally in terms of the real numbers. However, the geometrical counterpart of any *interval of numbers* is immediately available and is called an *interval of points*. Henceforth we shall make no serious effort to distinguish between sets of numbers and sets of points when the sense is clear. Unless there is good reason for careful distinction we shall feel free to use the words *number* and *point* interchangeably.

Definition.   *Let a and b be any two real numbers such that $a < b$. Then the* **open interval** *from a to b, written (a, b),† is the set of all numbers x between a and b: $a < x < b$. The* **closed interval** *from a to b, written $[a, b]$, is the set of all numbers x between a and b together with a and b: $a \leq x \leq b$. The* **half-open intervals** *$(a, b]$ and $[a, b)$ are defined by the inequalities $a < x \leq b$ and $a \leq x < b$, respectively. In each of these cases the interval is called a* **finite interval** *and the points a and b are called* **endpoints***, a being the* **left-hand endpoint** *and b being the* **right-hand endpoint***. Any point of an interval that is not an endpoint is called an* **interior point** *of the interval. The point $\frac{1}{2}(a + b)$ is the* **midpoint** *of each interval defined above.*

Example.   The point 3 is the left-hand endpoint (but not a member) of the open interval $(3, 10)$, the right-hand endpoint (and a member) of the closed interval $[-2, 3]$, and the midpoint of the half-open interval $(-2, 8]$.

803. INFINITY SYMBOLS. INFINITE INTERVALS

We introduce now two symbols for concepts that are useful adjuncts to the real number system.   These are $+\infty$, known as **plus infinity**, and $-\infty$, known as **minus infinity**.   These are related to the real numbers *by order alone*, and *not* by the algebraic operations of addition or multiplication.‡   The two infinities are related to the real numbers as follows: minus infinity is less than every real number, and every real number is less than plus infinity:

(1) $-\infty < x < +\infty$,   *for every real number x.*

The set consisting of all real numbers x and both infinities, subject to the inequalities (1), is called the **extended real number system**.   It is important to remember that $+\infty$ *and* $-\infty$ *are not numbers* (and therefore will not be referred to herein as *points*), and that *the extended real number system is not a field*, since addition and multiplication are not defined for all its members. (In fact, it is *impossible* to extend the operations of addition and multiplication to the extended real number system and retain the structure of a field—cf. Ex. 28, § 806.)

---

† Because of context, the use of parentheses for open intervals is not likely to be confused with their use for ordered pairs or GCD.   Other notations for the open interval (a, b) are: $]a, b[$ and $\langle a, b \rangle$.

‡ It is possible to define restricted operations among numbers and infinities and to use these in the theory of limits.   We shall have no occasion in this book to consider such combinations.   For a brief discussion of algebraic operations involving $+\infty$ and $-\infty$, see the author's *Advanced Calculus* (New York, Appleton-Century-Crofts, Inc., 1961), p. 42 (Exs. 43–56).

We are now ready to complete our catalog of intervals by adding to the family of finite intervals of § 802 the infinite intervals of the following definition:

Definition. *Let a be an arbitrary real number. Then the sets of real numbers denoted and defined as follows are* **infinite** *intervals having a as* **endpoint***:*

$$(a, +\infty): a < x < +\infty, \quad or \quad a < x;$$
$$[a, +\infty): a \leq x < +\infty, \quad or \quad a \leq x;$$
$$(-\infty, a): -\infty < x < a, \quad or \quad x < a;$$
$$(-\infty, a]: -\infty < x \leq a, \quad or \quad x \leq a.$$

The set $\mathcal{R}$ of all real numbers is an **infinite interval***, also denoted:*

$$(-\infty, +\infty): -\infty < x < +\infty, \quad or \quad x \in \mathcal{R}.$$

Any point of an interval that is not an endpoint of that interval is called an **interior point** *of the interval.*

804. ABSOLUTE VALUE. NEIGHBORHOODS

Definition I. *The* **absolute value** *of a number x, written $|x|$, is defined:*

$$|x| \equiv \begin{cases} x & if \ x \geq 0, \\ -x & if \ x < 0. \end{cases}$$

Example 1. $|5| = 5, |-3| = 3, |0| = 0.$

The absolute value of a number can be thought of as its (nonnegative) **distance** from the **origin** 0 in Figure 801. Similarly, the absolute value of the difference between two numbers, $|x - y|$, is the **distance** between the two points x and y. Some of the more useful properties of absolute value are given in the theorem:

Theorem I. Properties of Absolute Value

I. $|x| \geq 0$; $|x| = 0$ *if and only if $x = 0$.*

II. $|xy| = |x| \cdot |y|$.

III. $\left|\dfrac{x}{y}\right| = \dfrac{|x|}{|y|} (y \neq 0).$

IV. *If $\epsilon > 0$, the inequality $|x| < \epsilon$ is equivalent to the simultaneous inequalities $-\epsilon < x < \epsilon$.*

V. *If $\epsilon \geq 0$, the inequality $|x| \leq \epsilon$ is equivalent to the simultaneous inequalities $-\epsilon \leq x \leq \epsilon$.*

VI. *For all real x, $-|x| \leq x \leq |x|$.*

VII. *The* **triangle inequality**† *holds:* $|x + y| \leq |x| + |y|.$

† Property VII is called the *triangle inequality* because the corresponding inequalities for complex numbers and ror vectors state that any side of a triangle is less than or equal to the sum of the other two (equality holding only in case of a degenerate triangle).

VIII. $|-x| = |x|$; $|x - y| = |y - x|$.

IX. $|x|^2 = x^2$.

X. $|x - y| \leqq |x| + |y|$.

XI. $\big| |x| - |y| \big| \leqq |x - y|$.

Proof. I: If $x > 0$, $|x| > 0$; if $x < 0$, $|x| = -x > 0$; therefore if $x \neq 0$, $|x| > 0$. If $x = 0$, $|x| = 0$. II: If either x or y (or both) is zero, $|xy| = 0 = |x| \cdot |y|$. If x and y are both positive, $|xy| = xy = |x| \cdot |y|$. If x and y are both negative, $|xy| = xy = (-x)(-y) = |x| \cdot |y|$. If x and y have opposite signs with, say, $x > 0$ and $y < 0$, $|xy| = -xy = x(-y) = |x| \cdot |y|$. III: $|x| = |(x/y) \cdot y| = |x/y| \cdot |y|$, by II. Divide by $|y|$. IV: If $|x| < \epsilon$ and $x \geqq 0$, then $-\epsilon < 0 \leqq x < \epsilon$; if $|x| < \epsilon$ and $x < 0$, then $-x < \epsilon$, so that $-\epsilon < x < 0 < \epsilon$. Conversely, if $-\epsilon < x < \epsilon$ and $x \geqq 0$, then $|x| < \epsilon$; if $-\epsilon < x < \epsilon$ and $x < 0$, then $-x < \epsilon$ so that $|x| < \epsilon$. V: This is similar to IV. VI: This follows from V with $\epsilon = |x|$. VII: From V we have both $-|x| \leqq x \leqq |x|$ and $-|y| \leqq y \leqq |y|$, and hence, after addition, $-(|x| + |y|) \leqq (x + y) \leqq |x| + |y|$. The conclusion follows from V with $x + y$ and $|x| + |y|$ playing the roles of x and ϵ, respectively. VIII: If $x = 0$, $|-x| = 0 = |x|$. If $x > 0$, $-x < 0$ and $|-x| = -(-x) = x = |x|$. If $x < 0$, $-x > 0$ and $|-x| = -x = |x|$. $|x - y| = |-(y - x)| = |y - x|$. IX: If $x \geqq 0$, $|x|^2 = x^2$. If $x < 0$, $|x|^2 = (-x)^2 = x^2$. X: By the triangle inequality and VIII, $|x - y| = |x + (-y)| \leqq |x| + |-y| = |x| + |y|$. XI: By the triangle inequality, $|x| = |y + (x - y)| \leqq |y| + |x - y|$, and after subtraction of $|y|$, $|x| - |y| \leqq |x - y|$. Similarly, by VII and VIII, with the roles of x and y, interchanged, $|y| - |x| \leqq |y - x| = |x - y|$. Therefore, $-|x - y| \leqq |x| - |y| \leqq |x - y|$, and XI follows from V.

Definition II. *A* **neighborhood** *of a point (real number) a is an open interval of the form $(a - \epsilon, a + \epsilon)$, where ϵ is some positive number.*

Theorem II. *The neighborhood $(a - \epsilon, a + \epsilon)$ of the point a consists of all points x whose distance from a is less than ϵ: $|x - a| < \epsilon$. The point a is the midpoint of each of its neighborhoods.*

Proof. The neighborhood $(a - \epsilon, a + \epsilon)$ consists of all x such that $a - \epsilon < x < a + \epsilon$. This double inequality is equivalent, by subtraction of a, to $-\epsilon < x - a < \epsilon$ and hence, by IV, Theorem I, to the single inequality $|x - a| < \epsilon$. The endpoints of the interval $(a - \epsilon, a + \epsilon)$ are $a - \epsilon$ and $a + \epsilon$, and hence the midpoint is $\frac{1}{2}[(a - \epsilon) + (a + \epsilon)] = a$.

805. SOLVING INEQUALITIES†

If an inequality (in terms of any of the symbols $<$, $\leqq$, $>$, and $\geqq$) involves an unknown quantity represented, say, by the letter x, we say that a certain

† Cf. the references on inequalities given in the footnote of § 202.

value of x **satisfies** the inequality if and only if the inequality resulting from replacing x (wherever it occurs) by the particular value is a true statement. A **solution** of an inequality involving x is any value of x that satisfies the inequality. To **solve** an inequality involving x means to find all solutions; that is, *all* values of x that satisfy the inequality.

Example 1. The inequality $3x - 1 < x + 5$ is satisfied by $x = 1$ and is not satisfied by $x = 4$, as can be seen in both cases by direct substitution. To solve this inequality let us start by assuming that x satisfies it. Adding 1 to each member gives $3x < x + 6$; then, subtracting x tells us that $2x < 6$. Finally, division by the *positive* number 2 gives the answer: $x < 3$. At least, what we have done ensures that every solution of the original inequality is less than 3. To show that every number less than 3 satisfies the given inequality we have only to reverse the foregoing steps: Assuming that $x < 3$, we have, by multiplying by the positive number 2, $2x < 6$. Add x to both sides: $3x < x + 6$. Finally, subtraction of 1 from both sides of this inequality gives the original, $3x - 1 < x + 5$, and the conclusion that every number less than 3 is a solution. We infer, therefore, that the set of all solutions of the given inequality is the set of all x less than 3.

If we define two inequalities involving x to be **equivalent** if and only if they have the same solutions, we see easily that the operations typified in the solution of Example 1 transform an inequality into an equivalent inequality. Specifically:

Theorem. *If an inequality consisting of two members and involving an unknown x is transformed by any one of the following four operations the resulting inequality is equivalent to the original (the expressions $A(x)$, $B(x)$, $C(x)$, and $D(x)$ involving x are arbitrary except as restricted):*

   (i) *the same quantity $A(x)$ is added to both members;*
   (ii) *the same quantity $B(x)$ is subtracted from both members;*
   (iii) *both members are multiplied by the same positive quantity $C(x)$ (that is, $C(x) > 0$ for all x);*
   (iv) *both members are divided by the same positive quantity, $D(x)$ (that is, $D(x) > 0$ for all x).*

Proof. If x satisfies an inequality and if the same quantity is added to both members to produce a second inequality, then by Example 4, § 203, x must also satisfy the second inequality. On the other hand, if x satisfies the second inequality it must also satisfy the first, for similar reasons. In other words, part (*i*) of the theorem is established. Part (*ii*) is proved in like fashion except that the operations of addition and subtraction are interchanged. Finally, parts (*iii*) and (*iv*) follow in parallel fashion with the aid of Exercise 6, § 204.

Example 1 (again). Solve the inequality $3x - 1 < x + 5$ by use of the preceding theorem.

Solution. Each of the following inequalities is equivalent to each of its predecessors, and therefore the last one gives the solutions: $3x - 1 < x + 5$; $3x < x + 6$; $2x < 6$; $x < 3$.

Example 2. The inequality $\dfrac{x}{x^2 + 5} > \dfrac{2}{x^2 + 5}$ is equivalent to the inequality $x > 2$, since $x^2 + 5 > 0$ for all real x. The inequality $\dfrac{x}{x - 1} \geqq \dfrac{2}{x - 1}$ is not equivalent to $x \geqq 2$ since $x - 1$ is not always positive. (For example, $x = 0$ is a solution of the first inequality but not of the second.)

Example 3. Solve the inequality $\dfrac{x}{x - 1} \geqq \dfrac{2}{x - 1}$ of Example 2.

Solution. The inequality $\dfrac{x}{x - 1} \geqq \dfrac{2}{x - 1}$ is equivalent (by (*ii*) of the Theorem) to $\dfrac{x - 2}{x - 1} \geqq 0$, and is therefore satisfied if and only if either $x = 2$ or the two quantities $(x - 2)$ and $(x - 1)$ are nonzero and have the same sign. They are both positive if and only if $x > 2$, and they are both negative if and only if $x < 1$. Therefore x is a solution of the given inequality if and only if either $x \geqq 2$ or $x < 1$.

Example 4. Solve the inequality $\dfrac{x - 2}{x - 1} < \dfrac{x + 2}{x + 1}$.

Solution. A temptation for the inexperienced might be to "cross-multiply" and write $x^2 - x - 2 < x^2 + x - 2$, with $x > 0$ defining the solutions. Without bothering to see how grossly incorrect this procedure is, let us turn our attention to solving the given inequality correctly. The given inequality is equivalent to

$$\frac{x + 2}{x + 1} - \frac{x - 2}{x - 1} = \frac{2x}{(x + 1)(x - 1)} > 0.$$

This is true if and only if the three quantities x, $(x + 1)$, and $(x - 1)$ either are all positive or have one positive member and two negative members among them. Since these three quantities are equal to 0 at $x = 0$, $x = -1$, and $x = 1$, respectively, we have only to examine their signs in the four intervals $(-\infty, -1)$, $(-1, 0)$, $(0, 1)$, and $(1, +\infty)$. In the first interval the quantities $x + 1$, x, and $x - 1$ are all negative, in the second interval they have the signs $+$, $-$, and $-$, respectively, in the third interval they are $+$, $+$, and $-$, respectively, and in the fourth interval they are all positive. Therefore x satisfies the given inequality if and only if x is in either the second interval or the fourth interval, that is, if and only if either $-1 < x < 0$ or $x > 1$.

Example 5. Solve the inequality $|x - 3| > 2x + 1$.

Solution. We first seek solutions in the interval $(-\infty, 3)$. If $x < 3$, the given inequality is equivalent to $3 - x > 2x + 1$, or $x < \frac{2}{3}$. In other words, if $x < 3$, then the given inequality is satisfied if and only if $x < \frac{2}{3}$. Now for the interval $[3, +\infty)$: If $x \geqq 3$, the given inequality is equivalent to $x - 3 > 2x + 1$, or $x < -4$. The statement that if $x \geqq 3$ the given inequality is satisfied if and only if $x < -4$ means that it is *never* satisfied if $x \geqq 3$. In other words, the given inequality is satisfied if and only if $x < \frac{2}{3}$.

Example 6. Solve the inequality $|x - 3| > |2x + 1|$.

First solution. As in Example 5, this can be solved by considering separate intervals, in this case $(-\infty, -\frac{1}{2}]$, $(-\frac{1}{2}, 3)$ and $[3, +\infty)$. If $x \leq -\frac{1}{2}$ the given inequality is equivalent to $3 - x > -2x - 1$, or $x > -4$, and the solutions from the first interval are the points of the interval $(-4, -\frac{1}{2}]$. If $-\frac{1}{2} < x < 3$ the given inequality is equivalent to $3 - x > 2x + 1$, or $x < \frac{2}{3}$; the solutions lying in $(-\frac{1}{2}, 3)$ make up the interval $(-\frac{1}{2}, \frac{2}{3})$. If $x \geq 3$, the given inequality is equivalent to $x - 3 > 2x + 1$, or $x < -4$, and there are no solutions. Therefore the solutions of the problem are the points of the interval $(-4, \frac{2}{3})$: $-4 < x < \frac{2}{3}$.

Second solution. By Exercise 10, § 204, the given inequality is equivalent to $|x - 3|^2 > |2x + 1|^2$ or, by IX of Theorem I, § 804, to $x^2 - 6x + 9 > 4x^2 + 4x + 1$. This last inequality is equivalent to $3x^2 + 10x - 8 = (x + 4)(3x - 2) < 0$. The two factors $(x + 4)$ and $(3x - 2)$ have opposite signs if and only if $-4 < x < \frac{2}{3}$.

806. EXERCISES

All quantities appearing in these exercises are assumed to be real numbers (or, more generally, members of an ordered field $\mathcal{G}$).

1. Prove that $|x_1 x_2 \cdots x_n| = |x_1| \cdot |x_2| \cdots \cdots |x_n|$.

2. Prove the *general triangle inequality:*

$$|x_1 + x_2 + \cdots + x_n| \leq |x_1| + |x_2| + \cdots + |x_n|.$$

In Exercises 3–16, find the values of x that satisfy the given inequality, or inequalities. Express your answer without absolute values.

3. $|x - 2| < 3$.   **4.** $|x + 3| \geq 2$.

5. $|x - 5| < |x + 1|$.   **6.** $|x - 4| > x - 2$.

7. $|x - 4| \leq 2 - x$.   **8.** $|x - 2| > x - 4$.

9. $|x^2 - 5| \leq 4$.   **10.** $x^2 + 10 < 6x$.

11. $|x + 5| < 2|x|$.   **12.** $|x| > 2x + 3$.

13. $x < x^2 - 12 < 4x$.   **14.** $|x - 7| < 5 < |5x - 25|$.

15. $| |x| - 1 | > |x + 1|$.   **16.** $|x - 1| > |x| - 2$.

In Exercises 17–22, solve for x, and express your answer in a simple form by using absolute value signs.

17. $\dfrac{x - a}{x + a} > 0, a \neq 0$.   **18.** $\dfrac{a - x}{a + x} \geq 0, a \neq 0$.

19. $\dfrac{x^2}{x^2 - 4} \leq \dfrac{9}{x^2 - 4}$.   **20.** $\dfrac{x^4}{x^4 - 16} < \dfrac{5x^2 + 36}{x^4 - 16}$.

21. $\dfrac{x - a}{x - b} > \dfrac{x + b}{x + a}$.   **22.** $\dfrac{x - a}{x - b} > \dfrac{x + a}{x + b}$.

23. Replace by a single equivalent inequality:

$$x > a + b, \; x > a - b.$$

24. Prove that

$$\max (a, b) = \tfrac{1}{2}(a + b) + \tfrac{1}{2} |a - b|,$$
$$\min (a, b) = \tfrac{1}{2}(a + b) - \tfrac{1}{2} |a - b|.$$

(Cf. Example 4, § 306.)

25. Prove that if $a^n = b^n$, where n is a nonzero integer, then $|a| = |b|$.

26. If x and y are given numbers and if $|x - y| \leq \epsilon$ for every positive number ϵ, prove that $x = y$. (Cf. Ex. 13, § 204.)

27. Prove the following statements concerning the quadratic polynomial $ax^2 + bx + c$: If $ax^2 + bx + c \geq 0$ for every x, then either $a = 0$, $b = 0$, and $c \geq 0$, or $a > 0$, and $b^2 - 4ac \leq 0$, and conversely. If $ax^2 + bx + c > 0$ for every x, then either $a = 0$, $b = 0$, and $c > 0$, or $a > 0$ and $b^2 - 4ac < 0$, and conversely. *Hint:* If $a = 0$ and $b \neq 0$, show that the inequality $bx + c < 0$ has a solution. If $a \neq 0$, write $ax^2 + bx + c = a(x + b/2a)^2 + (c - b^2/4a)$. If $a < 0$ show that the inequality $(x + b/2a)^2 > (b^2 - 4ac)/4a^2$ has a solution. (Cf. Ex. 15, § 204.)

28. Let F and G be fields, with $F \subset G$, and assume that G contains exactly two members that are not members of F. Prove that G is finite, and is therefore not an ordered field. In fact, prove that G is isomorphic to the field of order four of Note 2, § 606. *Hint:* Let a and b be the two special members of G. Since a and b are not members of F, neither are $-a$ and $-b$. Consider the two cases: (*i*) $-a = a$ and $-b = b$, and (*ii*) $-a = b$ and $-b = a$, showing that case (*i*) leads to a field F consisting of only 0 and 1, and that case (*ii*) is impossible.

9

The Axiom of Completeness

◆◆◆

901. UPPER AND LOWER BOUNDS

The time has come to introduce the final axiom of the real number system. This axiom, called the *axiom of completeness* or the *axiom of continuity*, can be fashioned in many equivalent ways. In the main text of this book we have chosen one of these. In the Appendix seventeen other equivalent forms are listed, and the equivalence of all of these is established. (In the exercises of the Appendix, there are an additional seven statements for the student to prove equivalent to completeness.) The relation of order is fundamental throughout.

In this chapter it will be assumed, unless explicit statement to the contrary is made, that *all sets under consideration are sets of real numbers.* It should be immediately obvious that much of the discussion can be transferred to general ordered fields (and in many cases to totally ordered systems (cf. the Appendix, § 1) in general). It is for simplicity of language primarily that we restrict our consideration to sets of real numbers.

The axiom of completeness, as stated in § 902, is based on the idea of *upper bound.*

Definition I. *If A is any set of real numbers, a number u is called an* **upper bound** *of A if and only if a $\leq$ u for every member a of A. A set A is*

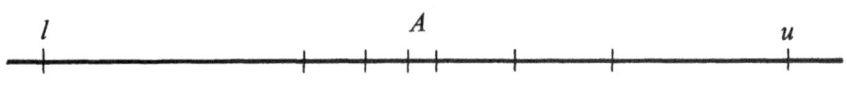

FIG. 901

bounded above *if and only if it has an upper bound. A number l is called* *a* **lower bound** *of A if and only if l $\leq$ a for every member a of A. A set A is* **bounded below** *if and only if it has a lower bound. A set is* **bounded** *if and only if it is bounded above and below. (Cf. Fig. 901.) A set is* **un-bounded above, unbounded below,** *or* **unbounded** *if and only if it fails to be bounded above, bounded below, or bounded, respectively.*

Examples. The set $\mathscr{R}$ of all real numbers is unbounded above, since there is no largest real number (cf. Theorem VII, § 312). For a similar reason, $\mathscr{R}$ is unbounded below. The set $\mathscr{P}$ of all positive numbers is also unbounded above, but $\mathscr{P}$ is bounded below (by 0 and by any negative number).

NOTE 1. The empty set $\varnothing$ is bounded since every real number is both an upper bound and a lower bound of $\varnothing$. (For example, if u is an arbitrary real number, u is an upper bound of $\varnothing$ since the only way in which the inequality $a \leq u$ could *fail* to hold for all members a of $\varnothing$ is for there to *exist* a member a of $\varnothing$ greater than u.)

NOTE 2. A set is bounded if and only if it is a subset of some finite interval. (Why?)

902. SUPREMA AND INFIMA

Whenever a set if bounded above it has many upper bounds, and whenever a set is bounded below it has many lower bounds. For example, if u is an upper bound of a set A then any number greater than u (such as $u + 1$) is also an upper bound of A, and if l is a lower bound of A so is any number less than l.

Suppose, now, that λ is an upper bound of a nonempty set A. Let us ask the question: "Is there an upper bound of A *less* than λ?" If the answer to this question is "No," then λ is the *least* of all possible upper bounds of A. We frame this concept formally in the definition:

Definition I. *Let A be a nonempty set that is bounded above, and assume that a number λ exists having the two following properties:*
(i) λ is an upper bound of A;
(ii) $\lambda \leq u$ for every upper bound u of A.
*Then λ is called the **least upper bound** or **supremum** of A, and is written.*†

$$(1) \qquad\qquad \lambda = \sup (A) = \sup A$$

NOTE 1. If the supremum of A exists it is unique, so that the definite article *the* of Definition I is appropriate. This can be seen by letting λ and μ designate two suprema of A, so that since each is an upper bound of A (by (*i*)) it follows from (*ii*) that $\lambda \leq \mu$ and $\mu \leq \lambda$, and therefore $\lambda = \mu$.

NOTE 2. If a number λ is the supremum of a set A, the set A is said to *have* the supremum λ, whether λ is a member of the set A or not.

Example 1. The number b is the supremum of the closed interval $[a, b]$ since it is an upper bound, and clearly no number *less* than b is an upper bound of the entire interval. (For example, no number c less than b can satisfy the inequality $c \geq x$ for all members x of $[a, b]$ since this inequality would fail for the particular member $x = b$.) The number b is also the supremum of the *open* interval (a, b) (without being a member of it). In this case, we see that no number c less than b can be an upper bound of (a, b) by assuming there exists such a c. Since $\frac{1}{2}(a + b) \in (a, b)$,

† The least upper bound of A is also denoted l.u.b. (A) and l.u.b. A.

$\frac{1}{2}(a + b) \leq c < b$, and $c \in (a, b)$. Therefore $c < \frac{1}{2}(c + b) \in (a, b)$, and c is not an upper bound after all!

Example 2.   Show that if a nonempty set A has a greatest member μ, then μ is the supremum of A:

(2) $$\mu = \max (A) = \sup (A).$$

Conclude that *every nonempty finite set has a supremum*, and that *any nonempty set possessing a supremum that is not a member of the set must be infinite*.

Solution.   Let μ be the greatest member of a set A. Then $\mu \geq a$ for every member a of A, and μ is an upper bound of A. If u is an upper bound of A, then $u \geq a$ for every $a \in A$ and, in particular, $u \geq \mu$. Therefore μ is the supremum of A, by (*i*) and (*ii*) of Definition I.   Since every nonempty set has a greatest member (Theorem VI, § 312) the two statements of the last sentence follow immediately.

The essential question now is whether a given set *has* a least upper bound, or supremum.   Of course, if a set is not bounded above it has no upper bound and *a fortiori* no *least* upper bound.   Suppose a nonempty set *is* bounded above.   Then does it have a least upper bound?   The answer is given by the last axiom:

V.   Axiom of Completeness.   *Every nonempty set of real numbers that is bounded above has a least upper bound.*

In general, any ordered field with the property just described is called *complete*, according to the definition:

Definition II.   *A complete ordered field is an ordered field in which every nonempty set that is bounded above has a least upper bound.*

In terms of this definition, the axiom of completeness for the real number system can be reduced to the statement that *the real number system is a complete ordered field.*

The axiom of completeness can be regarded geometrically as stating that there are no "gaps" in the real axis or number scale.   This idea is illustrated in the following example:

Example 3.   Let $\mathcal{R}$ be the real number system, and let x be any real number. Show that the set S obtained by deleting from $\mathcal{R}$ the single point x is not complete. That is, find a nonempty subset of S that is bounded above by a member of S and that has no least upper bound in S.

Solution.   Let A be the set of all numbers less than x.   Then A is nonempty ($x - 1$ is a member), and A is bounded above in S ($x + 1$ is a member of S that is an upper bound of A).   To show that A has no supremum in S, let y denote a member of S that is assumed to be such a least upper bound of A.   There are two cases.   If $y < x$, then (Ex. 14, § 204) $y < \frac{1}{2}(x + y) < x$, and y is less than $\frac{1}{2}(x + y)$, a member of A, and y is thus not even an upper bound of A.   If $y > x$, then $y > \frac{1}{2}(x + y) > x$, and $\frac{1}{2}(x + y)$ is an upper bound of A less than the assumed *least* upper bound y.

Let us now turn our inequalities around and look at a set from below.

Definition III. *Let A be a nonempty set that is bounded below, and assume that a number γ exists having the two following properties:*

(i) *γ is a lower bound of A,*

(ii) *γ ≧ l for every lower bound l of A.*

Then γ is called the greatest lower bound or infimum of A, and is written:†

$$(3) \qquad\qquad \gamma = \inf (A) = \inf A.$$

NOTE 3. If the infimum γ of a set A exists it is unique. In such an event A is said to *have* γ as an infimum whether γ belongs to A or not. If a set A has a least member ν, then ν is the infimum of A: $\nu = \min (A) = \inf (A)$. Every finite set has both a supremum and an infimum.

The property of the real number system that every nonempty set bounded above has a least upper bound has as companion the property that every nonempty set bounded below has a greatest lower bound. These two statements are equivalent to each other in any ordered field. If either is taken as an axiom the other can be proved as a theorem. (These and other similar matters are discussed at some length in the Appendix.) Since we have chosen to take the existence of suprema as our axiom, we shall now establish existence of infima as a theorem:

Theorem. *Every nonempty set of real numbers that is bounded below has a greatest lower bound.*

Proof. Let A be a given nonempty set of real numbers, and let B be the set consisting of every real number x such that $-x \in A$. (The set B is a "mirror image" of the set A, the "mirror" being located at the origin.) Then B is nonempty since A is nonempty. If y is any lower bound of A, then $-y$ is an upper bound of B. (The inequality $-y > x = -a$ is equivalent to the inequality $y < -x = a$.) Let λ denote the supremum of B: $\lambda \equiv \sup (B)$, and define $\gamma \equiv -\lambda$. We wish to prove that γ is the infimum of A: $\gamma = \inf (A)$. In the first place, γ is a lower bound of A. (The inequality $\gamma \le a$ is equivalent to the inequality $\lambda = -\gamma \ge -a$.) In the second place, if l is a lower bound of A, then $-l$ is an upper bound of B and hence, by (ii) of Definition I, $-l \ge \lambda = -\gamma$. From this we infer $l \le \gamma$ and thus property (ii) of Definition III, and the proof is complete.

Example 4. Show that the set of all lower bounds of a nonempty set A of real numbers is either empty or an interval of the form $(-\infty, c]$. (Cf. Ex. 1, § 908.)

Solution. Let B be the set of all lower bounds of A and assume that B is nonempty. If c is the infimum of A, we wish to show that $B = (-\infty, c]$. In the first place, every member of B is a lower bound of A and hence is less than or equal to the

---

† The greatest lower bound of A is also denoted g.l.b. (A) and g.l.b. A.

greatest lower bound c, so that $B \subset (-\infty, c]$. In the second place, every member x of $(-\infty, c]$ satisfies the inequality $x \leq c$ and is therefore a lower bound of A: $x \in B$, so that $(-\infty, c] \subset B$.

903. THE ARCHIMEDEAN PROPERTY

One of the most important consequences of the axiom of completeness (§ 902) is the algebraic counterpart of a basic principle of Euclidean geometry known as the *Archimedean property*.† This principle states that any length (however large) can be exceeded by repeatedly marking off a given length (however small), each successive application starting where the preceding one stopped. (A midget ruler, if used a sufficient number of times, can measure off an arbitrarily large distance.) For real numbers this principle, again called the Archimedean property, has the following formal statement and proof (cf. Theorem I, § 403; Note, § 501; Ex. 8, § 506; and Definition II, § 713):

Theorem I. Archimedean Property. *If a and b are positive numbers, there is a positive integer n such that $na > b$.*

Proof. If the theorem were false, the inequality $na \leq b$ would hold for *all* positive integers n. That is, the set $\{a, 2a, 3a, \cdots\}$ would be bounded above. Let c be the least upper bound of this set. Then $na \leq c$ for all n, and hence $(n + 1)a \leq c$ for all n. Therefore $na + a \leq c$, or $na \leq c - a$, for all n. Thus $c - a$ is an upper bound that is *less* than the *least* upper bound c. This is the desired contradiction.

Three corollaries are immediate:

Corollary I. *If x is any real number there exists a positive integer n such that $n > x$.*

Proof. If $x \leq 0$, let $n = 1$. If $x > 0$, use Theorem I with $a = 1$ and $b = x$.

Corollary II. *If ϵ is any positive number there exists a positive integer n such that*

$$\frac{1}{n} < \epsilon.$$

Proof. Let $n > 1/\epsilon$, by Corollary I.

Corollary III. *If x is any real number there exist integers m and n such that*

$$m < x < n.$$

† Cf. D. Hilbert, *The Foundations of Geometry* (La Salle, Ill., The Open Court Publishing Co., 1938).

Proof. By Corollary I, there exist positive integers n and p such that $n > x$ and $p > -x$. Let $m \equiv -p$.

With the aid of these results we can now establish another important relation between real numbers in general and integers in particular.

Theorem II. *If x is any real number there exists a unique integer n such that*
$$n \leq x < n + 1.$$

Proof. *Existence:* By Corollary III, Theorem I, there exist integers r and s such that $r < x < s$ and therefore, since $t = s - r$ is a positive integer, there also exists a positive integer t such that $r < x < r + t$. Let p be the *least* positive integer such that $r + p > x$. (Such a p exists by the well-ordering principle, VII, of § 304.) Finally, let $n \equiv r + p - 1$. If $p = 1$, then $n = r < x < r + 1 = n + 1$. If $p \geq 2$, then $p - 1$ is a positive integer and $n = r + (p - 1) \leq x$ by the minimum property of p. Hence $n \leq x < r + p = n + 1$. *Uniqueness:* If m and n are distinct integers such that $m \leq x < m + 1$ and $n \leq x < n + 1$, assume for definiteness that $m < n$. Then $m < n \leq x < m + 1$, and hence $0 < n - m < 1$, in contradiction of Property II, § 304, which states that there is no positive integer (natural number) less than 1. This completes the proof.

NOTE. The integer n of Theorem II, uniquely determined by the real number x, defines a real-valued function of the real variable x, denoted $[x]$. Since, for any real number x, the integer $[x]$ is the greatest integer less than or equal to x, this function is called the **greatest integer function**. It is also sometimes called the **bracket function**.

904. DENSITY OF THE RATIONAL NUMBERS

Definition. *A set S of real numbers contained in an interval I is **dense** in I if and only if between any two distinct points of I there is a member of S; that is, if a and b are members of I and $a < b$ there exists a member s of S such that*
$$a < s < b.$$

We now establish a fundamental relation between the real number system $\mathscr{R}$ and the rational number system $\mathscr{Q}$ (Chapter 5):

Theorem I. *The set $\mathscr{Q}$ of rational numbers is dense in $\mathscr{R}$. The set of rational numbers that are members of an interval I is dense in I.*

Proof. Let a and b be any two real numbers, where $a < b$, or $b - a > 0$. By Corollary II, Theorem I, § 903, there exists a positive integer q such that $\frac{1}{q} < b - a$. We now seek an integer p so that $\frac{p}{q}$ shall satisfy the inequalities
$$a < \frac{p}{q} \leq a + \frac{1}{q} < b.$$

This will hold if p is chosen so that $aq < p \leq aq + 1$, or $p \leq aq + 1 < p + 1$. Such a p exists by Theorem II, § 903.

It is of interest that between any two distinct real numbers there are *infinitely many* rational numbers. This is a consequence of the corresponding fact about dense sets in general.

Theorem II. *If S is dense in an interval I, and if a and b are any members of I with $a < b$, then there are infinitely many members of S between a and b.*

Proof. By definition, we know that the set A of all members of S between a and b is nonempty. Assume it is finite. Then, by Example 4, § 306, A has a greatest member λ, so that for every member x of A: $a < x. \leq \lambda < b$. But, since S is dense in I, there exists a member x of S between λ and b, and hence a member x of A between λ and b, in contradiction to the inequality $x \leq \lambda$, which holds for all x in A.

A consequence of the density of the rational numbers that will be important to us later is stated in the theorem:

Theorem III. *If x is any real number and if A is the set of all rational numbers less than x, then $x = \sup A$. If B is the set of all rational numbers greater than x, then $x = \inf B$.*

Proof. Only the first statement will be proved since the second is entirely similar, with inequalities reversed. In the first place, since x is an upper bound of A and $\sup A$ is the *least* upper bound, $\sup A \leq x$. Assume that $\sup A < x$. Then, by the density of the set $\mathcal{Q}$ of rational numbers (Theorem I), there exists a rational number r between $\sup A$ and x: $\sup A < r < x$. But this means that $\sup A$ is not even an upper bound of the set A since it is exceeded by a member r of A. With this contradiction the proof is complete.

905. EXISTENCE AND DENSITY OF THE IRRATIONAL NUMBERS

With the powerful tool of the axiom of completeness (§ 902) at our disposal we can now—at long last—establish the existence of irrational numbers. In Chapter 12 we shall demonstrate the existence of vast quantities of irrational numbers (Theorem II, § 1214). In the present chapter we shall content ourselves with a special infinite set of irrational numbers, all resting on a particular positive number known as **the square root of** 2, and denoted $\sqrt{2}$. (A thorough study of roots of numbers is presented in Chapter 10.)

As an aid in the proof of the theorem that follows, we establish a lemma:

Lemma. *If $0 < s < 1$ and if y is any real number, then*

(1) $$(y + s)^2 < y^2 + s(y + 1)^2,$$

(2) $$(y - s)^2 > y^2 - s(y + 1)^2.$$

Proof. Inequality (1) is equivalent to $2sy + s^2 < sy^2 + 2sy + s$, which is equivalent to $s < y^2 + 1$. Inequality (2) is equivalent to $-2sy + s^2 > -sy^2 - 2sy - s$, which is equivalent to $s > -y^2 - 1$.

Theorem I. *There exists a positive number whose square is equal to 2.*

Proof. Assume there does *not* exist a positive number whose square is equal to 2, let A be the set of all positive numbers a whose squares are *less* than 2: $a^2 < 2$, and let B be the set of all positive numbers b whose squares are *greater* than 2: $b^2 > 2$. The set $\mathscr{P}$ of positive numbers is thus divided into two sets A and B having the following three properties: (*i*) *A is nonempty* $(1 \in A)$; (*ii*) *B is nonempty* $(2 \in B)$; (*iii*) *if* $a \in A$ *and* $b \in B$ *then* $a < b$ (if $0 < b \leq a$ then by Exercise 10, § 204, $0 < 2 < b^2 \leq a^2 < 2$). Therefore A, being a nonempty bounded set, has a least upper bound y: $y \equiv \sup A$. We shall obtain our desired contradiction by showing that *the positive number y is neither a member of A nor a member of B.* Before proceeding with the details of the proof we first observe that $1 < y < 2$. This is true for the following two reasons: In the first place $\frac{4}{3} \in A$ and hence (since y is an upper bound of A) $y \geq \frac{4}{3} > 1$. In the second place, $\frac{3}{2} \in B$ and hence (since every member of B is an upper bound of A and y is the *least* upper bound of A) $y \leq \frac{3}{2} < 2$.

We start the main part of the proof by assuming that y is a member of A: $y^2 < 2$. The contradiction sought in this case is that *there is a member of A greater than y* (and therefore that y is not an upper bound of A after all). The idea is to find a positive number s so small that the positive number $y + s$, greater than y, is also a member of A: $(y + s)^2 < 2$. Such a positive number s can be found by means of inequality (1) of the Lemma, if we require s to be a positive number less than 1 such that $y^2 + s(y + 1)^2 < 2$. Since y^2 is assumed to be less than 2, such a positive number s is given by the formula:

$$(3) \qquad\qquad s \equiv \tfrac{1}{2} \min \left(\frac{2 - y^2}{(y + 1)^2}, 1 \right),$$

since (3) guarantees simultaneously the inequalities $0 < s < 1$ and $s < (2 - y^2)/(y + 1)^2$.

Finally, we assume that y is a member of B: $y^2 > 2$. The contradiction now sought is that *there is a member b of B less than y* (and therefore that y is not the *least* upper bound of A). In the present case the idea is to find a positive number s so small that the positive number $y - s$, less than y, is also a member of B: $(y - s)^2 > 2$. Such a positive number s can be found by means of inequality (2) of the Lemma, if we require s to be a positive number less than 1 (and hence less than y) such that $y^2 - s(y + 1)^2 > 2$. Since y^2 is assumed to be greater than 2, such a positive number s is given by the formula:

$$(4) \qquad\qquad s \equiv \tfrac{1}{2} \min \left(\frac{y^2 - 2}{(y + 1)^2}, 1 \right),$$

since (4) guarantees simultaneously the inequalities $0 < s < 1$ and $s < (y^2 - 2)/(y + 1)^2$. This completes the proof of Theorem I.

Theorem II. *The positive number $\sqrt{2}$ whose square is equal to 2 is unique.*

Proof. Assume that there are two distinct positive numbers, x and y, with $x < y$, such that $x^2 = y^2 = 2$. By Exercise 10, § 204, $x^2 < y^2$. (Contradiction.)

Theorem III. *The number $\sqrt{2}$ is irrational. That is, there do not exist integers p and q such that $p/q = \sqrt{2}$.*

Proof. If there existed integers p and q such that $p/q = \sqrt{2}$, then there would exist *positive* integers p and q such that $p/q = \sqrt{2}$. Furthermore, by Theorem II, § 402, there would exist *relatively prime* positive integers p and q such that $p/q = \sqrt{2}$, or $p^2 = 2q^2$. By Theorem IV, § 404, since $q \mid p^2$ and $(p, q) = 1$, $q \mid p$ and hence $q = 1$. But this means that $p = \sqrt{2}$ is a positive integer between 1 and 2. With this contradiction the proof is complete. (Cf. Ex. 21, § 710, Exs. 5–6, § 1010.)

As a consequence of Theorem III it is now easy to demonstrate that the ordered field of rational numbers fails to have the important property of completeness enjoyed by the real number system.

Theorem IV. *The ordered field $\mathscr{Q}$ of rational numbers is not complete.*

Proof. Assume that $\mathscr{Q}$ is complete and consider the set A of all rational numbers r such that $r < \sqrt{2}$. Then A is nonempty ($1 \in A$), and A is bounded above by a member of $\mathscr{Q}$ (if $r \in A$ then $r < 2$). Therefore A has a least upper bound c in $\mathscr{Q}$, by assumption. Since $\sqrt{2}$ is irrational, there are only two possibilities for c. If $c < \sqrt{2}$, then by the density of the rationals, there is a rational number a such that $c < a < \sqrt{2}$; that is, there is a member of A greater than c, and c is not an upper bound of A. If $c > \sqrt{2}$, then there is a rational number b such that $\sqrt{2} < b < c$; that is, there is a member of $\mathscr{Q}$ that is an upper bound of A less than the *least* upper bound c. These two contradictions complete the proof.

NOTE 1. It might be claimed that a new proof of Theorem IV is unnecessary since, by Theorem III, § 904, if A is the set of all rationals less than $\sqrt{2}$, sup $A = \sqrt{2}$, which is irrational. However, the supremum occurring in Theorem III, § 904, is relative to the ordered field $\mathscr{R}$ of real numbers while that of the present theorem is relative to a different ordered field, that of the rational numbers only. To illustrate the idea with an example, consider the set S consisting of all points of the half-open interval [0, 1) together with the two points 2 and 3, and let T be the subset got by deleting from S the single point 2. If A is the interval [0, 1), then in S the supremum of A is 2, while in T the supremum of A is 3.   The fact that 2 is not a member of T does not imply that T fails to be complete.   As it happens, T *is* complete.   (Why?)

Before establishing the density of the set of irrational numbers we prove a useful pair of theorems, and one corollary:

Theorem V.   *If r is a nonzero rational number and if x is irrational, then the following numbers are all irrational:* $x + r,\ x - r,\ r - x,\ xr,\ x/r,\ r/x.$

Proof.   Each part of the proof is by contradiction and uses the fact that the system $\mathcal{Q}$ of rational numbers is a field (§ 504).   Two sample demonstrations should be sufficient; the rest are left for the reader to supply.   Assume that $x + r$ is a rational number s: $x + r = s$.   Then $x = r - s$, which is rational (contradiction).   Assume that r/x is a rational number s: $r/x = s$.   Then $s \neq 0$ and $x = r/s$, which is rational (contradiction).

Corollary.   *Every number of the form $\sqrt{2}r$, where r is a nonzero rational number, is irrational.*

Theorem VI.   *The system of irrational numbers is neither closed under addition nor closed under multiplication, and consequently is no one of the following algebraic entities: (i) additive group (Ex. 30, § 104); (ii) multiplicative group (Ex. 31, § 104); (iii) ring (§ 606), (iv) integral domain (§ 501), (v) field (§ 102).*

Proof.   By Theorems III and V, the numbers $\sqrt{2}$ and $-\sqrt{2}$ are irrational; but their sum and their product are rational.

Theorem VII.   *The set S of irrational numbers is dense in $\mathcal{R}$.*

Proof.   Let x and y be any two distinct real numbers, with $x < y$.   Then $\sqrt{2}x < \sqrt{2}y$ and there exists a nonzero rational number r such that $\sqrt{2}x < r < \sqrt{2}y$ (if $\sqrt{2}x < 0 < \sqrt{2}y$, choose r so that $0 < r < \sqrt{2}y$).   Therefore,

$$x < \frac{r}{\sqrt{2}} < y,$$

and $r/\sqrt{2}$ is irrational, by Theorem V.

906. AN ADDED ITEM OF SET NOTATION

It will be convenient in the remainder of this book to have available a useful notation for sets defined by a certain property, or by certain properties. This notation† has the form $\{\cdots \mid \cdots\}$, where the symbol for the general member of the set being defined is written between the first brace $\{$ and the vertical bar $\mid$, and where the property or properties that define the set are written between the vertical bar $\mid$ and the second brace $\}$.   In the following illustrative examples, $\mathcal{R}$ represents the real number system, $\mathcal{Q}$ the ordered

† This is sometimes called the **set-builder** method.

field of rational numbers of $\mathcal{R}$, $\mathcal{I}$ the integral domain of integers of $\mathcal{R}$, $\mathcal{N}$ the set of natural numbers of $\mathcal{R}$, and $\mathcal{P}$ the set of positive numbers of $\mathcal{R}$.

Example 1. $\{x \mid x \in \mathcal{R}, x > 0\} = \mathcal{P}$.

Example 2. $\{x \mid x \in \mathcal{R}, x \leq 3\} = (-\infty, 3]$.

Example 3. $\{2n \mid n \in \mathcal{I}\}$ is the set of all even integers.

Example 4. If $A \subset \mathcal{R}$ and $B \subset \mathcal{R}$, then $\{a + b \mid a \in A, b \in B\}$ is the set of all real numbers that can be obtained by adding a member of A and a member of B.

Example 5. $\{n^2 \mid n \in \mathcal{N}\}$ is the set of all numbers that are squares of natural numbers.

Example 6. For a given real number x, $\{y \mid y \in \mathcal{R}, y - x \in \mathcal{Q}\}$ is the set of all real numbers y such that $y - x$ is rational. This same set can be written in the form $\{x + z \mid z \in \mathcal{Q}\}$.

907. CATEGORICAL NATURE OF THE AXIOMS

It was shown in § 606 that for every prime number p there exists a finite field having exactly p members. This fact shows that the axioms of a field are not sufficient to determine the exact nature of that field—the field may consist of 5 members, 13 members, or infinitely many members.

The axioms of an ordered field are more restrictive—for example, they rule out all finite fields (Theorem V, § 312)—but it is still possible to find quite distinct examples of ordered fields. We have, indeed, three such examples: (*i*) the ordered field $\mathcal{Q}$ of rational numbers, (*ii*) the ordered field $\mathcal{R}$ of real numbers, and (*iii*) the ordered field $\mathcal{H}$ of rational functions from $\mathcal{R}$ to $\mathcal{R}$. These three ordered fields are *essentially distinct* in the sense that, for any two of the three, one of the two has a fundamental property not shared by the other. The ordered field (*iii*) of rational functions, for instance, is the only one of the three that is non-Archimedean; and (*i*) and (*ii*) differ in that the equation $x^2 = 2$ has a solution in (*ii*) but not in (*i*). (For another important ordered field distinct from these three, see the Note, § 1214. Also cf. Ex. 4, § 908.)

Suppose we add the axiom of completeness. Then how many essentially distinct complete ordered fields are there? The answer is simple and unequivocal: "One." It is our purpose in this section to explain what "essentially distinct" really means, and to prove that all complete ordered fields are structurally the same. In other words, there is no ambiguity in speaking of *the* real number system, since any two "real number systems" are essentially the same.

For present purposes we restate the essential features of the Definition, § 311:

Definition. *Two ordered fields $\mathcal{G}$ and $\mathcal{G}'$ are* **isomorphic** *if and only if their members can be put into a one-to-one correspondence preserving both* *operations of addition and multiplication, as well as preserving the order relation; that is, if and only if there exists a one-to-one correspondence denoted* $x \leftrightarrow x'$, *where* $x \in \mathcal{G}$ *and* $x' \in \mathcal{G}'$, *such that the two correspondences*

$$x \leftrightarrow x' \quad and \quad y \leftrightarrow y'$$

imply the two correspondences

$$x + y \leftrightarrow x' + y' \quad and \quad xy \leftrightarrow x'y',$$

or, in another form,

$$(x + y)' = x' + y' \quad and \quad (xy)' = x'y',$$

and, furthermore,

$$x < y \quad if \ and \ only \ if \quad x' < y'.$$

We now come to our principal theorem:

Theorem. *Any two complete ordered fields are isomorphic.*

Proof. The first parts of the proof have already been given in §§ 311, 502, and 505. In § 311 it was shown that if $\mathcal{G}$ and $\mathcal{G}'$ are any two ordered fields, then the systems of natural numbers, $\mathcal{N}$ of $\mathcal{G}$ and $\mathcal{N}'$ of $\mathcal{G}'$, are isomorphic. In § 502 the isomorphism between $\mathcal{N}$ and $\mathcal{N}'$ is extended to the integral domains of integers, $\mathcal{I}$ of $\mathcal{G}$ and $\mathcal{I}'$ of $\mathcal{G}'$. In § 505 this isomorphism is extended once more to the ordered fields of rational numbers, $\mathcal{Q}$ of $\mathcal{G}$ and $\mathcal{Q}'$ of $\mathcal{G}'$. In the present section we shall extend the isomorphism one final time to the complete ordered fields $\mathcal{R}$ and $\mathcal{R}'$ themselves. For this last extension, of course, the axiom of completeness must play a key role. The manner in which the completeness property appears is in the density of the rational numbers and, in particular, in the special kind of density behavior described in Theorem III, § 904, wherein it is shown that any member of a complete ordered field is the supremum of the set of all rational numbers less than it.

For an arbitrary member x of the complete ordered field $\mathcal{R}$ let A be the set of all rational numbers of $\mathcal{R}$ less than x: $A = \{r \mid r \in \mathcal{Q}, r < x\}$. Then $x = \sup (A)$. Let A' be the set of rational numbers of $\mathcal{R}'$ that correspond to the members of A according to the isomorphic correspondence already established for the rational number fields $\mathcal{Q}$ and $\mathcal{Q}'$ (that is $A' = \{r' \mid r \in A, r \leftrightarrow r'\}$), and define x' to be the least upper bound of the set A' (which is bounded above in $\mathcal{R}'$ (why?)): $x' \equiv \sup (A')$. We note in passing that if x happens to be a rational number, then the correspondence $x \leftrightarrow x'$ agrees with the one already in existence. (Why?)

The correspondence $\sup (A) \leftrightarrow \sup (A')$ is one-to-one and order-preserving between $\mathcal{R}$ and $\mathcal{R}'$, as we shall now show in two parts. In the first place, we wish to show that the inequality $x < y$ (where x and y are members of $\mathcal{R}$) implies the inequality $x' < y'$. Let r and s be rational numbers in $\mathcal{R}$ such that $x < r < s < y$, let A be the set of rational numbers of $\mathcal{R}$ less than x, and let B be the set of rational numbers less than y. If A' and B' are the sets of rational numbers in $\mathscr{R}'$ that correspond to the members of A and B, respectively, then, since r' is an upper bound of A', $x' = \sup(A') \leq r'$, and since $s' \in B'$ and $y' = \sup(B')$, $s' \leq y'$. Therefore $x' \leq r' < s' \leq y'$, and $x' < y'$ as desired. In the second place, if ξ is any member of $\mathscr{R}'$, we wish to show that there exists a member x of $\mathscr{R}$ such that $x' = \xi$. Let C' be the set of all rational numbers in $\mathscr{R}'$ that are less than ξ and let C be the set of all rational numbers in $\mathscr{R}$ to which the members of C' correspond. Then C is bounded above in $\mathscr{R}$ and $x \equiv \sup(C)$ exists. If A is the set of all rational numbers in $\mathscr{R}$ that are less than x, then $x = \sup(A)$, and we wish to prove that if $x' \equiv \sup(A')$, then $x' = \xi$. Since C' has no greatest member, C has no greatest member, so that every member of C is less than x, $C \subset A$, $C' \subset A'$, and $\xi \leq x'$. If $\xi < x'$, let r' be a rational number in $\mathscr{R}'$ such that $\xi < r' < x'$, and let r be the rational number of $\mathscr{R}$ to which r' corresponds. Since r' is an upper bound of the set C', r is an upper bound of C. On the other hand, $r < x$ since the equality $r = x$ implies the equality $r' = x'$ and the inequality $r > x$ implies the inequality $r' > x'$. This means that, in $\mathscr{R}$, $\sup(C) \leq r < x = \sup(C)$, and a contradiction has been reached. Therefore $x' = \xi$, as desired. One form of expressing this result is: $(\sup A)' = \sup(A')$. With the fact established that for any x' in $\mathscr{R}'$ there exists an x in $\mathscr{R}$ to which x' corresponds, we are now in a position (since $x \geq y$ implies $x' \geq y'$) to state that the inequality $x' < y'$ in $\mathscr{R}'$ implies the inequality $x < y$ in $\mathscr{R}$.

As an aid in the remaining two major parts of the proof we now show that if x is an arbitrary member of $\mathscr{R}$, if A is the set of all rational numbers in $\mathscr{R}$ less than x: $A = \{r \mid r \in \mathscr{Q}, r < x\}$, and if x' and A' correspond to x and A, respectively, in $\mathscr{R}'$, then A' is the set of all rational numbers in $\mathscr{R}'$ that are less than x': $A' = \{r' \mid r' \in \mathscr{Q}', r' < x'\}$. In the first place, since every member of A is a rational number in $\mathscr{R}$ that is less than x, every member of A' is a rational number in $\mathscr{R}'$ that is less than x'. In other words, A' consists of at least *some* of the rational numbers in $\mathscr{R}'$ that are less than x'. If there were a rational number s' *not* in A' such that $s' < x'$, then $s < x$ in $\mathscr{R}$. That is, $s \in A$, and hence $s' \in A'$. (Contradiction.)

In order to show that $(x + y)' = x' + y'$, we let $A \equiv \{r \mid r \in \mathscr{Q}, r < x\}$, $B \equiv \{s \mid s \in \mathscr{Q}, s < y\}$, $C \equiv \{t \mid t \in \mathscr{Q}, t < x + y\}$, and $A + B \equiv \{a + b \mid a \in A, b \in B\}$. The first thing to show is that $C = A + B$. Since for every a in A and b in B, $a < x$ and $b < y$ it follows that $a + b < x + y$, and hence $A + B \subset C$. On the other hand, if c is an arbitrary rational number less than $x + y$: $c < x + y$, so that $c - y < x$, let a be any rational number such that $c - y < a < x$. (In particular, $a \in A$.) If $b \equiv c - a$, then $b = c - a < y$ so that $b \in B$, and $a + b = c$ so that c is a member of $A + B$, or $C \subset A + B$. From the equation $C = A + B$ it follows that $x + y = \sup(C) = \sup(A + B)$. Using primes as before, $A' = \{r' \mid r' \in \mathscr{Q}', r' < x'\}$, $B' = \{s' \mid s' \in \mathscr{Q}', s' < y'\}$, $C' = \{t' \mid t' \in \mathscr{Q}', t' < (x + y)'\}$, and if $A' + B' \equiv \{a' + b' \mid a' \in A', b' \in B'\}$, then by the isomorphism $r \leftrightarrow r'$ between $\mathscr{Q}$ and $\mathscr{Q}'$,

$C' = (A + B)' = A' + B'$. By the same reasoning as that used immediately above in the proof that $C = A + B$ we obtain the fact that $A' + B'$ is the set of all rational numbers less than $x' + y'$. The equality of the two sets C' and $A' + B'$ now implies the equality of their suprema, and we conclude $(x + y)' = x' + y'$, as desired.

Before proceeding to the last part of the proof we draw an inference from what has just been established: For any x in $\mathscr{R}$, $-(-x)' = x'$. To show this we let $y \equiv -x$, so that $x + y = 0$. Therefore, since $0'$ is the zero element of $\mathscr{R}'$ and since $(x + y)' = x' + y' = 0'$, we have $x' = -y' = -(-x)'$, as desired.

Finally, we wish to show that $(xy)' = x'y'$. If either x or y is zero the proof is trivial. For the first portion of this part of the proof we *assume that x and y are both positive*, and let $A \equiv \{r \mid r \in \mathscr{Q}, 0 < r < x\}$, $B \equiv \{s \mid s \in \mathscr{Q}, 0 < s < y\}$, $C \equiv \{t \mid t \in \mathscr{Q}, 0 < t < xy\}$, and $AB \equiv \{ab \mid a \in A, b \in B\}$. The first thing to show is that $C = AB$. The inclusion $AB \subset C$ is trivial. On the other hand, if c is an arbitrary positive rational number less than xy:

$0 < c < xy$, then $0 < \dfrac{c}{y} < x$. Let a be an arbitrary rational number such that $\dfrac{c}{y} < a < x$. (In particular, $a \in A$.) If $b = \dfrac{c}{a}$, then $0 < b = \dfrac{c}{a} < y$ so that $b \in B$, and $ab = c$ so that c is a member of AB, or $C \subset AB$. From the equation $C = AB$ it follows that $xy = \sup(C) = \sup(AB)$. The details now parallel those of the proof that $(x + y)' = x' + y'$. In outline: $C' = (AB)' = A'B'$, and the set of all positive rationals less than $(xy)'$ is the same as the set of all positive rationals less than $x'y'$. Hence $(xy)' = x'y'$, if x and y are both positive. Using the remarks of the preceding paragraph we can complete the proof as follows: If x and y are both negative, $(xy)' = ((-x)(-y))' = (-x)'(-y)' = (-(-x)')(-(-y)') = x'y'$. If $x < 0$ and $y > 0$, $(xy)' = -(-xy)' = -((-x)y)' = -((-x)'y') = (-(-x)')y' = x'y'$, and similarly if $x > 0$ and $y < 0$. This completes all details of the proof.

NOTE. Any specific example satisfying all axioms of an axiomatic system is called a **model** of that system. For example, the integers modulo 7 is a model for a field. An axiomatic system is said to be **categorical** if and only if every two models of it are isomorphic. The Theorem of this section, then, states that *Axioms I through V for the real number system are categorical*. For a discussion of categoricalness and the related concept of *completeness*, with particular relation to axiomatic systems for geometry, see R. L. Wilder, *The Foundations of Mathematics* (New York, John Wiley and Sons, Inc., 1952), pp. 36–40.

908. EXERCISES

All numbers indicated in these exercises are assumed to be real numbers, that is, members of the complete ordered field $\mathscr{R}$.

1. Prove that the set of all upper bounds of a nonempty set of numbers is either empty or an interval of the form $[c, +\infty)$.

2. Prove that any set resulting from the deletion of a finite number of points from a dense set of numbers is dense.

3. If a is an integer greater than 1, prove that the set of all numbers of the form p/a^q, where p is an integer and q is a positive integer, is dense in $\mathscr{R}$. (For further discussion of such numbers, see Chapter 12.)

4. Prove that any number of the form $r + s\sqrt{2}$, where r and s are rational, has a unique representation in this form. Prove that the set of all such numbers constitutes an ordered field. (Cf. Ex. 7, § 1010.)

5. Let S be an arbitrary bounded nonempty set of real numbers. Prove there is a *smallest* closed interval I containing S, in the following sense: If J is any closed interval containing S, then J contains I.

6. Prove by counterexample that the statement of Exercise 5 is false if the word *closed* is replaced by the word *open*.

7. Let S be a nonempty set of numbers bounded above, and let x be the least upper bound of S. Prove that x has the two properties corresponding to an arbitrary positive number ϵ: (*i*) every element s of S satisfies the inequality $s < x + \epsilon$; (*ii*) at least one element s of S satisfies the inequality $s > x - \epsilon$.

8. Prove that the two properties of Exercise 7 characterize the least upper bound. That is, prove that a number x subject to these two properties is the least upper bound of S.

9. State and prove the analogue of Exercise 7 for greatest lower bounds.

10. State and prove the analogue of Exercise 8 for greatest lower bounds.

11. Let x be an arbitrary irrational number, and let a, b, c, and d denote rational numbers such that c and d are not both zero. Prove that $(ax + b)/(cx + d)$ is rational if and only if $ad = bc$. Be sure to consider all possible cases.

12. Let A and B be nonempty sets of numbers such that $A \subset B$. If A and B are bounded above, prove that sup $A \leqq$ sup B. If A and B are bounded below, prove that inf $A \geqq$ inf B. In each case give an example to show that a strict inequality may occur.

13. If A is any nonempty set of numbers, let $-A$ denote the set of all numbers of the form $-a$, where $a \in A$. Prove that if A is bounded above then

$$\text{sup } (A) + \text{inf } (-A) = 0,$$

and that if A is bounded below, then

$$\text{inf } (A) + \text{sup } (-A) = 0.$$

14. If A and B are arbitrary nonempty sets and if c is any number, the sets $A + B$, AB, $c + A$, and cA are defined as follows: $A + B$ is the set of all numbers of the form $a + b$, where $a \in A$ and $b \in B$; AB is the set of all numbers of the form ab, where $a \in A$ and $b \in B$; $c + A$ is the set of all numbers of the form $c + a$, where

$a \in A$; and cA is the set of all numbers of the ca, where $a \in A$. If A and B are bounded above, if c is any number, and if p is any positive number, prove:

$$\sup (A + B) = \sup A + \sup B,$$
$$\sup (c + A) = c + \sup A,$$
$$\sup (pA) = p \sup A.$$

If A and B are nonempty sets of nonnegative numbers that are bounded above, prove:

$$\sup (AB) = \sup A \cdot \sup B.$$

15. State and prove the analogue of Exercise 14 for infima.

16. Let S be an arbitrary nonempty set, and let f and g be functions on S into $\mathscr{R}$. Then f if **bounded above (below)** on S if and only if the range of f is bounded above (below). In case f is bounded above on S, $\sup (f) = \sup f$ is defined to be supremum of its range, with a similar definition for $\inf (f) = \inf f$. If f is bounded (that is, bounded above and below) on S, prove (cf. Note 3, § 310):

$$\sup f + \inf (-f) = 0,$$
$$\inf f + \sup (-f) = 0.$$

17. If f and g are functions on a nonempty set S into $\mathscr{R}$, if f and g are bounded above, if c is any constant, and if p is any positive constant, prove (cf. Note 3, § 310):

$$\sup (f + g) \leqq \sup f + \sup g,$$
$$\sup (c + f) = c + \sup f,$$
$$\sup (pf) = p \sup f.$$

In addition, if the values of f and g are nonnegative, prove:

$$\sup (fg) \leqq \sup f \cdot \sup g.$$

In the first and last case above, give an example to show that a strict inequality may occur.   (Cf. Ex. 16.)

18. State and prove the analogue of Exercise 17 for infima.

19. If f is a function on a nonempty set S into $\mathscr{R}$, the function $h = |f|$ is defined by the equation $h(x) = |f(x)|$. If f is a bounded function (cf. Ex. 16) on S into $\mathscr{R}$, define the **norm** of f: $\|f\| = \sup |f|$. Prove that the norm, for bounded functions f and g and constants c, has the properties:

(i) $\|f\| \geqq 0$; $\|f\| = 0$ if and only if $f = 0$;

(ii) $\|cf\| = |c| \cdot \|f\|$;

(iii) $\|f + g\| \leqq \|f\| + \|g\|$.

Show by an example that the inequality in (iii) may be a strict inequality.

20. Prove that any additive group in the real number system $\mathscr{R}$ consisting of more than one member either (i) has a smallest positive member a and consists of all integral multiples na of a, or (ii) is dense in $\mathscr{R}$.

21. A subset of an ordered field is said to be an **ordered subfield** if and only if the operations of addition and multiplication and the relation of order in the subset are consistent with those of the containing ordered field. Prove that every ordered subfield of an Archimedean ordered field is Archimedean.

22. In any Archimedean ordered field prove that the rational numbers are dense.

23. In any Archimedean ordered field prove that every number is the supremum of the set of all rational numbers less than it.   (Cf. Ex. 22.)

24. Prove that every Archimedean ordered field is isomorphic to an ordered subfield of the real number system; in other words, that every Archimedean ordered field can be isomorphically embedded in the real number system.   (Cf. Exs. 22, 23.)

10

Roots and Rational Exponents

◆◆◆

1001. INTRODUCTION

In order to establish the existence of irrational numbers, in Chapter 9, we showed that there is a positive number (denoted $\sqrt{2}$) whose square is equal to 2. In the present chapter we shall extend this result to arbitrary positive nth roots of positive numbers. This will make it possible to study powers of positive numbers for arbitrary rational exponents. Finally, in Chapter 11, exponentiation will be extended to arbitrary real exponents.

To facilitate the existence proofs of the following section, we prove two lemmas. These lemmas are suggested by a fairly simple application of the binomial theorem (Ex. 27, § 318), and can be proved by means of that theorem. We present, however, independent proofs.

Lemma 1. *If $0 < s < 1$ and $y > 0$, and if n is a natural number, then*

(1) $$(y + s)^n < y^n + s(y + 1)^n.$$

Proof. We shall establish (1) by proving the stronger result:

(2) $$(y + s)^n \leq y^n - sy^n + s(y + 1)^n.$$

We prove (2) by induction. If $P(n)$ is the statement that (2) is true, then $P(1)$ is a trivial equality. Assuming $P(n)$, we wish to prove $P(n + 1)$:

(3) $$(y + s)(y + s)^n \leq y^{n+1} - sy^{n+1} + s(y + 1)^{n+1}.$$

Making use of (2) in the second factor on the left of (3) reduces our problem to the inequality (to be proved):

(4) $$(y + s)[y^n - sy^n + s(y + 1)^n] \leq y^{n+1} - sy^{n+1} + s(y + 1)^{n+1}.$$

If the left-hand side of (4) is expanded to $y^{n+1} - sy^{n+1} + s(y^n - sy^n) + s(y + s)(y + 1)^n$, subtraction of two terms and division by s leads to the following inequality, equivalent to (4):

(5) $$y^n - sy^n + (y + s)(y + 1)^n \leq (y + 1)^{n+1},$$

or, equivalently:

(6) $$y^n(1 - s) \leq (y + 1)^n[(y + 1) - (y + s)] = (y + 1)^n(1 - s).$$

135

Since $y^n < (y+1)^n$ and $1-s>0$, we have shown that $P(n)$ implies $P(n+1)$. By the Fundamental Theorem of Mathematical Induction (Theorem III, § 303), the proof of (2), and hence of (1), is complete.

Lemma 2. *If $0 < s < 1$ and $0 < s < y$, and if n is a natural number, then*

$$(7) \qquad\qquad (y-s)^n > y^n - s(y+1)^n.$$

Proof. If $z \equiv y - s$, then, by Lemma 1, $(z+s)^n < z^n + s(z+1)^n$, and hence:

$$(8) \qquad\qquad y^n < (y-s)^n + s(y-s+1)^n.$$

Since $0 < s < 1$, then $0 < y < y - s + 1 < y + 1$, and $s(y-s+1)^n < s(y+1)^n$, and (7) follows.

1002. ROOTS OF NONNEGATIVE NUMBERS

We now have the machinery for establishing the basic theorem on roots of nonnegative numbers:

Theorem. *If x is a nonnegative number and if n is a natural number, there exists a unique nonnegative number y such that $y^n = x$, denoted:*

$$(1) \qquad\qquad y = \sqrt[n]{x}.$$

Proof. Uniqueness. If $0 \le y < z$, then $y^n < z^n$, and hence the equality $y^n = z^n$ implies the equality $y = z$ if y and z are nonnegative. (Cf. Ex. 8, § 308.)

Existence, $n = 1$. Let $y = x$.
Existence, $n > 1$, $x = 0$. Let $y = 0$.
Existence, $n > 1$, $x = 1$. Let $y = 1$.
Existence, $n > 1$, $x > 1$. Let A be the set of all positive numbers a such that $a^n < x$, let B be the set of all positive numbers b such that $b^n > x$, and assume that there does *not* exist a positive number z such that $z^n = x$. Then every positive number must be a member of either A or B, and every member of A is less than every member of B. The set A is nonempty ($1 \in A$) and bounded above (x is an upper bound since if $a \in A$, then $a^n < x < x^n$ and hence $a < x$). Therefore A has a positive least upper bound. Denote this positive least upper bound by y:

$$y \equiv \sup (A).$$

By assumption, either $y^n < x$ or $y^n > x$. We shall show that each of these two inequalities leads to a contradiction, and hence shall be forced to the conclusion: $y^n = x$. (*i*) Assume $y^n < x$, and let s be defined to be the positive number:

$$s \equiv \tfrac{1}{2} \min \left(1, \frac{x - y^n}{(y+1)^n}\right).$$

Then $0 < s < 1$ and therefore, by Lemma 1, § 1001, and the definition of s:

$$(y + s)^n < y^n + \frac{x - y^n}{(y + 1)^n} (y + 1)^n = x.$$

But this means that $y + s$ is a member of A greater than the least upper bound y of A. (Contradiction.) (*ii*) Assume $y^n > x$, and let s be defined to be the positive number:

$$s \equiv \tfrac{1}{2} \min \left(1, y, \frac{y^n - x}{(y + 1)^n} \right).$$

Then $0 < s < 1$ and $0 < s < y$ and therefore, by Lemma 2, § 1001:

$$(y - s)^n > y^n - \frac{y^n - x}{(y + 1)^n} (y + 1)^n = x.$$

But this means that $y - s$ is an upper bound for the set A less than its *least* upper bound y. (Contradiction.)

Existence, $n > 1$, $0 < x < 1$. By the preceding proof, since $1/x > 1$, there exists a positive number u such that $u^n = 1/x$. If $y \equiv 1/u$, then $y^n = x$.

1003. MONOTONIC FUNCTIONS

In § 316, the concepts of decreasing and strictly decreasing sequences were introduced. In this section we shall extend these notions to apply to more general functions than sequences, and adapt them to increasing functions as well as to decreasing functions.

Definition. *Let f be a real-valued function of a real variable whose domain of definition contains a set A consisting of more than one point. Then f is an* **increasing**† *function on A if and only if* $f(x_1) \le f(x_2)$ *whenever* x_1 *and* x_2 *are points of A such that* $x_1 < x_2$; *f is a* **strictly increasing** *function on A if and only if* $f(x_1) < f(x_2)$ *whenever* x_1 *and* x_2 *are points of A such that* $x_1 < x_2$. *The function f is* **decreasing**‡ *on A or* **strictly decreasing** *on A if and only if the inequality* $x_1 < x_2$ (x_1 *and* x_2 *in A*) *implies* $f(x_1) \ge f(x_2)$ *or* $f(x_1) > f(x_2)$, *respectively. The function f is* **monotonic** *on A if and only if it is either an increasing function on A or a decreasing function on A; f is* **strictly monotonic** *on A if and only if it is either strictly increasing on A or strictly decreasing on A. Deletion of the words "on A" in any of these formulations means that the set A is the entire domain of definition of the function f.*

Example 1. A constant function is both increasing and decreasing, and is therefore monotonic.

Example 2. The greatest integer function $[x]$ (Note, § 903) is an increasing function.

---

† Also called **monotonically increasing**, or **nondecreasing**.
‡ Also called **monotonically decreasing**, or **nonincreasing**.

Example 3. The polynomial $x^2 - 6x + 13$ is strictly increasing on the interval $[3, +\infty)$ and strictly decreasing on the interval $(-\infty, 3]$, as is seen by the form $x^2 - 6x + 13 = (x - 3)^2 + 4$.

Theorem. *Let n be a natural number and let A be the infinite interval $[0, +\infty)$ consisting of all nonnegative numbers x. Then each of the functions (i) x^n and (ii) $\sqrt[n]{x}$, with domain of definition A, is strictly increasing.*

Proof. (*i*): This is a consequence of Exercise 8, § 308. (*ii*): Assume that $0 \leq x_1 < x_2$ and let $y_1 \equiv \sqrt[n]{x_1}$ and $y_2 \equiv \sqrt[n]{x_2}$. If $y_1 \geq y_2$, then (by (*i*)) $y_1{}^n \geq y_2{}^n$. But this means that $x_1 \geq x_2$, in contradiction to the assumption $x_1 < x_2$. Therefore $y_1 < y_2$.

1004. CONTINUITY OF MONOTONIC FUNCTIONS

The property of *continuity* defined below can be formulated in much greater generality than appears here.† However, we have restricted our considerations to monotonic functions defined on dense subsets of intervals since these are the only kinds of functions that will be of interest to us in the remainder of this book (except for the Appendix), and since the formulation of continuity for monotonic functions is particularly simple.

Definition. *Let I be an interval of real numbers and let f be an increasing real-valued function whose domain A is dense in I. Then f is **continuous** at a point c of A if and only if* ‡

(*i*) $$f(c) = \sup \{f(x) \mid x \in A, x < c\},$$

(*ii*) $$f(c) = \inf \{f(x) \mid x \in A, x > c\},$$

*with the understanding that if c is the left-hand endpoint of I then only (ii) is relevant and if c is the right-hand endpoint of I then only (i) is relevant. A similar definition, with interchange of supremum and infimum in (i) and (ii), applies if f is a decreasing function. A function is **discontinuous** at any point where it is not defined or where it is defined and fails to be continuous.*

Example 1. The function $f(x) \equiv x$, defined on $\mathcal{R}$, is everywhere continuous since, for any real number c,

$$f(c) = c = \sup \{x \mid x < c\},$$
$$f(c) = c = \inf \{x \mid x > c\}.$$

Example 2. The greatest integer function $f(x) \equiv [x]$ (Note, § 903) is discontinuous at any integral value of x, $x = n$, since $[n] = n$, whereas

$$\sup \{[x] \mid x < n\} = n - 1 \neq n.$$

† Cf. the first footnote of § 3 of the appendix.
‡ Cf. § 906 for the $\{\cdots \mid \cdots\}$ notation.

Theorem. *Let n be a natural number and let A be the infinite interval $[0, +\infty)$ consisting of all nonnegative numbers x. Then each of the functions (i) x^n and (ii) $\sqrt[n]{x}$ with domain of definition A is continuous and unbounded. That is, each is continuous at every point of A, and the range of values of each is an unbounded set of numbers.*

Proof. (i): Continuity of x^n at $x = 0$ means that $0 = \inf \{x^n \mid x > 0\}$. Obviously, 0 is a lower bound of the set $\{x^n \mid x > 0\}$, and therefore $0 \leq \inf \{x^n \mid x > 0\}$. Assume that $0 < \gamma \equiv \inf \{x^n \mid x > 0\}$. Then $\gamma \leq x^n$ for every positive number x. But this is certainly false if $x = \frac{1}{2}\sqrt[n]{\gamma} < \sqrt[n]{\gamma}$. (Contradiction.) Continuity of x^n at $c > 0$ means that $c^n = \inf \{x^n \mid x > c\}$ $= \sup \{x^n \mid 0 \leq x < c\}$. Since x^n is an increasing function, $\sup \{x^n \mid 0 \leq x < c\} \leq c^n \leq \inf \{x^n \mid x > c\}$. Assume first that $\sup \{x^n \mid 0 \leq x < c\} = \sigma < c^n$. Then $0 \leq x < c$ implies $x^n \leq \sigma < c^n$, or (since $\sqrt[n]{x}$ is strictly increasing for $x \geq 0$) $x \leq \sqrt[n]{\sigma} < c$. A contradiction is obtained by taking x between $\sqrt[n]{\sigma}$ and c. Finally, assume that $c^n < \mu \equiv \inf \{x^n \mid x > c\}$. Then $x > c$ implies $c^n < \mu \leq x^n$, or $c < \sqrt[n]{\mu} \leq x$. If $c < x < \sqrt[n]{\mu}$ a contradiction results. To prove that x^n is unbounded, assume there exists a positive number b such that $x^n \leq b$ for all $x \geq 0$. This is clearly false for $x = \sqrt[n]{b} + 1$.

(ii): The details for this part are so closely parallel to those for part (i) that they are left as an exercise for the reader.

1005. RATIONAL EXPONENTS

Laws of exponents for all integral exponents and nonzero bases were proved in § 503. In the present section we shall extend the definition of power of a number to apply to an *arbitrary rational exponent* and *positive base*. In subsequent sections the laws of exponents will be proved, and properties of powers, considered both as functions of the exponent and as functions of the base, will be investigated.

Definition. *Let a be a positive number, and let x be a rational number p/q, where p and q are integers and q is positive. Then a^x is defined:*

$$(1) \qquad a^x = a^{\frac{p}{q}} \equiv \sqrt[q]{a^p}.$$

Theorem I. *The value of (1) is independent of the form of x. That is, if $x = p/q = m/n$, where $p, q, m,$ and n are integers and q and n are positive, then $\sqrt[q]{a^p} = \sqrt[n]{a^m}$.*

Proof. By the Theorem, § 1003, the equation $\sqrt[q]{a^p} = \sqrt[n]{a^m}$ is equivalent to $[\sqrt[q]{a^p}]^{nq} = [\sqrt[n]{a^m}]^{nq}$, and by laws of exponents for integral exponents the two members of this equation can be rewritten: $\{[\sqrt[q]{a^p}]^q\}^n = (a^p)^n = a^{np}$ and

$\{[\sqrt[n]{a^m}]^n\}^q = (a^m)^q = a^{mq}$, respectively. Finally, since $p/q = m/n$, $np = mq$ and the desired equality is obtained.

Theorem II. *Under the assumptions of the preceding Definition,*

(2)
$$\left(a^{\frac{p}{q}}\right)^q = a^p,$$

(3)
$$a^{\frac{p}{q}} = \sqrt[q]{a^p} = (\sqrt[q]{a})^p.$$

Proof. Equation (2) results from taking qth powers of the second and third quantities of (1). To prove (3) we observe that the equation $\sqrt[q]{a^p} = (\sqrt[q]{a})^p$ is equivalent to $[\sqrt[q]{a^p}]^q = [(\sqrt[q]{a})^p]^q$ and that this is equivalent to $a^p = [(\sqrt[q]{a})^q]^p = a^p$.

1006. LAWS OF EXPONENTS

Theorem. *If a and b are any positive numbers and if r and s are any rational numbers, then*

(i)
$$a^r a^s = a^{r+s};$$

(ii)
$$\frac{a^r}{a^s} = a^{r-s};$$

(iii)
$$(a^r)^s = a^{rs};$$

(iv)
$$(ab)^r = a^r b^r;$$

(v)
$$\left(\frac{a}{b}\right)^r = \frac{a^r}{b^r}.$$

Proof. (i): Let r and s be expressed as fractions with a positive common denominator: $r = m/q$, $s = n/q$, where m, n, and q are integers and $q > 0$ The equation to be established,

(1)
$$a^{\frac{m}{q}} a^{\frac{n}{q}} = a^{\frac{m+n}{q}},$$

is equivalent to $(a^{m/q}a^{n/q})^q = (a^{(m+n)/q})^q$, and this is equivalent by § 307 and Theorem II, § 1005, to $a^m a^n = a^{m+n}$, which is true by § 503.

(ii): By (i), $a^{r-s}a^s = a^{(r-s)+s} = a^r$.

(iii): Let $r = p/q$ and $s = m/n$, where p, q, m, and n are integers and q and n are positive. We wish to show:

(2)
$$\left(a^{\frac{p}{q}}\right)^{\frac{m}{n}} = a^{\frac{pm}{qn}}$$

or, equivalently, $[(a^{p/q})^{m/n}]^{qn} = [a^{pm/qn}]^{qn}$. This reduces to $(a^{p/q})^{mq} = a^{pm}$, by (2), § 1005, and this is true for the same reason.

(iv): The equation $(ab)^{p/q} = a^{p/q}b^{p/q}$ is equivalent to $(ab)^p = a^p b^p$, by the same principles as those used above, and this last equation is true by § 503.

(v):   By *(iv)*: $\left(\dfrac{a}{b}\right)^r b^r = \left(\dfrac{a}{b} \, b\right)^r = a^r.$

1007. THE FUNCTION a^x

If a is a fixed positive base and if x is an arbitrary rational number, then the expression a^x becomes a real-valued function with domain of definition the set $\mathcal{Q}$ of all rational numbers. This function is called the **exponential function** with **base** a, for rational values of the variable. The quantity x is called the **exponent**, and each value a^x is called a **power** of a. It is our immediate objective to study the exponential function, for rational values of the variable, with particular attention to its monotonic and continuity properties. We consider first the case $a > 1$.

For some of the details in the proofs of Theorems I and II the following lemma will be helpful (cf. Ex. 21, § 308).

Lemma. *If n is a natural number and if $\beta > 0$, then $(1 + \beta)^n \geqq 1 + n\beta$.*

Proof. If $P(n)$ is the statement of the lemma, then $P(1)$ is trivially true. Assuming $P(n)$ to be true, we find that $(1 + \beta)^{n+1} \geqq (1 + \beta)(1 + n\beta) = 1 + (n + 1)\beta + \beta^2 > 1 + (n + 1)\beta$. In other words, $P(n)$ implies $P(n + 1)$ and by the Fundamental Theorem of Mathematical Induction the lemma is proved.

Theorem I. *If $a > 1$ and if x is a variable restricted to rational values, the function a^x is strictly increasing on the set $\mathcal{Q}$ of all rational numbers and continuous at every rational point x. The values of the function a^x are all positive, and can be made arbitrarily large for positive x and less than an arbitrary positive number for negative x.*

Proof. To show that a^x is strictly increasing, we wish to show that $r < s$ implies $a^r < a^s$, where r and s are rational. Let r and s be represented by fractions having a positive common denominator: $r = m/q$, $s = n/q$, where m, n, and q are integers and $q > 0$. Then $r < s$ implies $m < n$, and we are to show that $\sqrt[q]{a^m} < \sqrt[q]{a^n}$ or, equivalently, that $a^m < a^n$. This is true since $a^{n-m} > 1$.

Since $a > 0$, $a^p > 0$ for every integer p, and hence $a^x = \sqrt[q]{a^p} > 0$. If b is any given number, we can apply the preceding Lemma to find a positive integer n such that $a^n > b$. In fact, if we let $\beta \equiv a - 1 > 0$ and take $n > (b - 1)/\beta$, then (by the preceding Lemma) $a^n = (1 + \beta)^n \geqq 1 + n\beta > 1 + [(b - 1)/\beta]\beta = b$. If ϵ is any given positive number, a negative integer n can be found such that $a^n < \epsilon$ as follows: let m be a positive integer such that $a^m > 1/\epsilon$, and let $n \equiv -m$.

Now let r be an arbitrary rational number. In order to establish continuity of a^x at $x = r$ we wish to show that $a^r = \sup\{a^x \mid x \in \mathscr{D}, x < r\} = \inf\{a^x \mid x \in \mathscr{D}, x > r\}$. With $x < r$, $a^r \geq \sup\{a^x \mid x \in \mathscr{D}, x < r\}$. If $a^r > \lambda \equiv \sup\{a^x \mid x \in \mathscr{D}, x < r\}$, then for every natural number n, $a^r > \lambda \geq a^{r-(1/n)}$, or $1 > \lambda/a^r \geq a^{-1/n}$ (by (ii) of the Theorem, § 1006). Therefore $\sqrt[n]{a} \geq a^r/\lambda > 1$ or, if $\beta \equiv (a^r/\lambda) - 1 > 0$, $a \geq (1 + \beta)^n \geq 1 + n\beta$ (by the preceding Lemma) for every natural number n. But the inequality $n\beta \leq a - 1$, for all natural numbers n, is forbidden by the Archimedean property (§ 903), and the desired contradiction is reached. Finally, $x > r$ implies $a^x > a^r$, and hence $a^r \leq \inf\{a^x \mid x \in \mathscr{D}, x > r\}$. If $a^r < \gamma \equiv \inf\{a^x \mid x \in \mathscr{D}, x > r\}$, then for every natural number n, $a^r < \gamma \leq a^{r+(1/n)}$, or $1 < \gamma/a^r \leq a^{1/n}$. Therefore, if $\beta \equiv (\gamma/a^r) - 1$, $a \geq (1 + \beta)^n \geq 1 + n\beta$ for all natural numbers n, and a contradiction is again obtained, as above. The proof of Theorem I is thus complete.

The proof of the following theorem is so similar to that of Theorem I that it is left as an exercise for the reader.

Theorem II. *If $0 < a < 1$ and if x is a variable restricted to rational values, the function a^x is strictly decreasing on the set $\mathscr{D}$ of all rational numbers and continuous at every rational point x. The values of the function a^x are all positive, and can be made arbitrarily large for negative x and less than an arbitrary positive number for positive x.*

1008. THE FUNCTION x^r

The expression x^r has been defined for all positive x and all rational r. For a fixed r, x^r becomes a function of the base x, called a **power function** of x. If r is *positive* the domain of this function can be extended to include 0 by defining 0^r to be 0, a result consistent with the equation $a^{p/q} = \sqrt[q]{a^p}$ of (1), § 1005, with p and q positive integers. For certain other values of r the domain of x^r can be extended to the entire real number system (cf. Ex. 4, § 1010, for specifications). In this section we shall state and prove a theorem concerning the function x^r for fixed $r > 0$ and variable $x \geq 0$, and state without proof a corresponding theorem for fixed $r < 0$ and variable $x > 0$.

Theorem I. *If r is a positive rational number and if x is a variable restricted to nonnegative values, the function x^r is strictly increasing on the interval $I = [0, +\infty)$ and continuous at every point x of I. The values of x^r are nonnegative and unbounded, and include $0^r = 0$ and $1^r = 1$.*

Proof. Let $r = p/q$, where p and q are both positive integers. If $0 \leq x_1 < x_2$, then $0 \leq x_1{}^p < x_2{}^p$ and $0 \leq \sqrt[q]{x_1{}^p} < \sqrt[q]{x_2{}^p}$, and hence $0 \leq x_1{}^r < x_2{}^r$. Continuity at $x = 0$ means that $0 = \inf\{x^r \mid x > 0\}$. Clearly $0 \leq \inf\{x^r \mid x > 0\}$. Assume $0 < \gamma \equiv \inf\{x^r \mid x > 0\}$. This means that for all positive numbers x, $x^{p/q} \geq \gamma$, or $x \geq \gamma^{q/p} > 0$. (Contradiction.) If c is any positive number,

$$\sup \{x^r \mid 0 \leq x < c\} \leq c^r \leq \inf \{x^r \mid x > c\}.$$

If $c^r > \sigma \equiv \sup \{x^r \mid 0 \leq x < c\}$, then $0 \leq x < c$ implies $x^{p/q} \leq \sigma < c^{p/q}$, or $x \leq \sigma^{q/p} < c$. (Contradiction.) If $c^r < \gamma \equiv \inf \{x^r \mid x > c\}$, then $x > c$ implies $x^{p/q} \geq \gamma > c^{p/q}$, or $x \geq \gamma^{q/p} > c$. (Contradiction.) Therefore continuity is established. Finally, to prove that x^r is unbounded, assume that $x^r \leq b$ for all $x \geq 0$. But this means that $x \leq b^{q/p}$ for all $x \geq 0$. With this last contradiction, the proof is complete.

Theorem II. *If r is a negative rational number and if x is a variable restricted to positive values, the function x^r is strictly decreasing on the interval $I = (0, +\infty)$ and continuous at every point x of I. The values of x^r are positive, and can be made arbitrarily large for small positive x and less than an arbitrary positive number for large positive x. The values of x^r include $1^r = 1$.*

Proof. The proof is left as an exercise for the reader.

*1009. AN INTEGRAL DOMAIN WITHOUT UNIQUE FACTORIZATION

In this section we give an example of an integral domain where the property of unique factorization (§ 405) does not obtain. (Cf. the references to the Birkhoff and MacLane book cited in the first paragraph of § 405.) We start with a definition concerning integral domains in general (cf. § 501).

Definition I. *A member u of an integral domain I is a **unit** of I if and only if it has a reciprocal in I; that is, if and only if there exists a member v of I such that $uv = 1$. A member z of I is **composite** if and only if it can be expressed as the product $z = xy$ of two nonzero members x and y of I neither of which is a unit. A nonzero member of I is **prime** if and only if it is neither a unit nor composite.*

Example 1. In the integral domain $\mathscr{I}$ of integers the units are 1 and -1.

Example 2. In any field every nonzero element is a unit.

Example 3. In an arbitrary integral domain, prove that no unit is composite. In fact, prove that if u is a unit, and if $u = xy$, then both x and y are units.

Solution. Assume that $u = xy$, and let v be such that $uv = 1$. Then $uv = x(yv) = y(xv) = 1$.

The example of an integral domain to be studied now will be denoted Φ and consists of all real numbers of the form $a + b\sqrt{5}$; where a and b are arbitrary integers. It should be noted that the representation of a number x in the form $x = a + b\sqrt{5}$ is unique since, if $a + b\sqrt{5} = c + d\sqrt{5}$ then $(a - c) = (d - b)\sqrt{5}$. If $d - b \neq 0$, then the irrational number $\sqrt{5}$ (cf. Ex. 5, § 1010) is equal to the rational number $(a - c)/(d - b)$. Therefore $b = d$, and consequently $a = c$. Since $(a + b\sqrt{5}) + (c + d\sqrt{5}) = (a + c) + (b + d)\sqrt{5}$ and $(a + b\sqrt{5})(c + d\sqrt{5}) = (ac + 5bd) + (ad + bc)\sqrt{5}$, Φ is closed with respect to the operations of addition and multiplication. It is easy to see, therefore, that Φ is an integral domain, and that Φ contains the integers $\mathscr{I}$ as a subdomain. For the integral domain Φ we introduce a special definition.

Definition II. *If $x = a + b\sqrt{5} \in \Phi$, then the **norm** of x, denoted $N(x)$ is the integer:*

$$(1) \qquad\qquad N(x) \equiv a^2 - 5b^2.$$

Theorem I. *(i) $N(x) = 0$ if and only if $x = 0$. (ii) $N(xy) = N(x)\,N(y)$. (iii) x is a unit if and only if $|N(x)| = 1$.*

Proof. If $x = 0$, then by the uniqueness of the representation $x = a + b\sqrt{5}$, $a = b = 0$ and $N(x) = 0$. If $N(x) = 0$ and $x \neq 0$, then $a - b\sqrt{5} = 0$ and $b \neq 0$, whence $\sqrt{5} = a/b$, in contradiction to the irrationality of $\sqrt{5}$. Verification of (ii) is routine substitution and is left to the reader. From (ii), if x is a unit and if y is the member of Φ such that $xy = 1$, then $|N(x)| \cdot |N(y)| = N(1) = 1$, and since $|N(x)|$ and $|N(y)|$ are natural numbers they must both be equal to 1. Conversely, if $x = a + b\sqrt{5} \in \Phi$ and if $N(x) = a^2 - 5b^2 = 1$ (or -1) then $y = a - b\sqrt{5}$ (or $y = -a + b\sqrt{5}$) is such that $xy = 1$ and x is a unit.

Theorem II. *In the integral domain Φ, the least value for the absolute value of the norm of a nonzero member that is not a unit is 4: if $x \in \Phi$, if $x \neq 0$, and if $|N(x)| \neq 1$, then $|N(x)| \geq 4$. Therefore, if $x \in \Phi$ and if x is composite, then $|N(x)| \geq 16$; equivalently, if $x \in \Phi$ and if $1 < |N(x)| < 16$, then x is prime.*

Proof. The problem is to show that the equation $a^2 - 5b^2 = c$, where $c = 2, -2, 3,$ or -3, has no solution in integers a and b. Assume that such a solution does exist, and write $a = 5q + r$, where q is an integer and $r = 0, 1, 2, 3,$ or 4, and write $c = 5n + s$, where $n = 0$ or -1, and $s = 2$ or 3. After substituting these quantities in the equation $a^2 - 5b^2 = c$ we see that 5 divides $r^2 - s$. But the only possible values for $r^2 - s$ are $-2, -1, 2, 7,$ and 14, if $s = 2$, and $-3, -2, 1, 6,$ and 13, if $s = 3$. Since none of these ten numbers are divisible by 5 we have obtained the desired contradiction.

In § 405, the Unique Factorization Theorem states that a natural number can be written as a product of prime natural numbers in one and, except for order of the factors, only one way. For the system $\mathscr{I}$ of integers the uniqueness is relaxed slightly to permit any factor to be multiplied by either 1 or -1. In a more general integral domain **unique factorization** is said to hold in case any factorization into prime factors is unique except for *order* and *multiplication of the factors by units*.

Theorem III. *In the integral domain Φ unique factorization fails. In fact, the following four things are all possible:* (i) *It is possible that a member x of Φ may be written in the form $x = p_1 p_2$ and in the form $x = q_1 q_2$, so that*

$$(2) \qquad\qquad p_1 p_2 = q_1 q_2,$$

where p_1, p_2, q_1, and q_2 are prime and neither factor on either side of (2) *is a unit times a factor on the other side.* (ii) *A prime member of Φ may divide the product of two members of Φ without dividing either factor separately.* (iii) *It may be that a fraction cannot be reduced to lowest terms in any essentially unique sense: two fractions a/b and c/d may be equal and in lowest terms and such that neither a nor c divides the other and neither b nor d divides the other.* (iv) *The concept of* GCD *may be meaningless: two members of Φ may not have a common divisor that is a multiple of every other common divisor.*

Proof. (i): By Theorem II, the numbers 2, $1 + \sqrt{5}$, and $-1 + \sqrt{5}$ are all prime since their norms are all equal to ± 4 and $1 < 4 < 16$. Equation (2), with $p_1 = p_2 = 2$, $q_1 = 1 + \sqrt{5}$, and $q_2 = -1 + \sqrt{5}$ is $2 \cdot 2 = (1 + \sqrt{5}) \times (-1 + \sqrt{5})$. Since $1 + \sqrt{5}$ cannot be written in the form $2(a + b\sqrt{5}) = 2a + 2b\sqrt{5}$, 2 is not a divisor of $1 + \sqrt{5}$. Therefore $1 + \sqrt{5}$ is not equal to a unit times 2. On the other hand, 2 is not equal to a unit times $1 + \sqrt{5}$, since if $2 = u(1 + \sqrt{5})$ and $uv = 1$, then $2v = 1 + \sqrt{5}$. (Contradiction.) A similar analysis with 2 and $-1 + \sqrt{5}$ completes this part of the proof.

(ii): 2 is not a divisor of either $1 + \sqrt{5}$ or $-1 + \sqrt{5}$, but it divides their product.

(iii): The fraction $2(1 + \sqrt{5})/4$ can be reduced to lowest terms in the following two ways:

$$\frac{2(1 + \sqrt{5})}{2 \cdot 2} = \frac{2(1 + \sqrt{5})}{(-1 + \sqrt{5})(1 + \sqrt{5})} = \frac{1 + \sqrt{5}}{2} = \frac{2}{-1 + \sqrt{5}},$$

With $a = 1 + \sqrt{5}$, $b = 2$, $c = 2$, and $d = -1 + \sqrt{5}$ we see from the proof of (i) that the statements of (iii) are true.

(iv): Let us assume that a "GCD" g of the two numbers 4 and $2(1 + \sqrt{5})$ exists in the sense that g divides both 4 and $2(1 + \sqrt{5})$ and, furthermore, that every common divisor of 4 and $2(1 + \sqrt{5})$ divides g. Since both 2 and $1 + \sqrt{5}$ are common divisors of 4 and $2(1 + \sqrt{5})$, then $2 \mid g$ and $(1 + \sqrt{5}) \mid g$, and g has the form $g = 2h$ and the form $g = (1 + \sqrt{5})w$. Since, by assumption, $2h \mid 4$ and $2h \mid 2(1 + \sqrt{5})$, and since the law of cancellation holds in Φ, it follows that $h \mid 2$ and $h \mid (1 + \sqrt{5})$, and we can write $2 = hx$ and $1 + \sqrt{5} = hy$. Taking norms, we have $4 = N(h)N(x)$ and $-4 = N(h)N(y)$. By Theorem II, $|N(h)| = 4$ or $|N(h)| = 1$. If $|N(h)| = 4$, then $|N(x)| = |N(y)| = 1$ and both x and y are units. But this means that $1 + \sqrt{5} = hy = 2(x^{-1}y)$.

This has already been shown to be impossible, and therefore $|N(h)| = 1$, and h is a unit. From $g = 2h = (1 + \sqrt{5})w$ we have $4N(h) = -4N(w)$, whence $|N(w)| = 1$, and w is also a unit. Therefore we can again write $1 + \sqrt{5}$ as a multiple of 2, thus: $1 + \sqrt{5} = 2(hw^{-1})$. This final contradiction completes the proof.

1010. EXERCISES

In these exercises $\mathscr{R}$ represents the real number system, and all numbers under consideration are members of $\mathscr{R}$.

1. If n is an odd positive integer, prove that the function x^n is strictly increasing on $\mathscr{R}$.

2. If n is an odd positive integer and x is any real number, prove that there exists one and only one real number y such that $y^n = x$.

3. If n is an odd positive integer, define and discuss the function $\sqrt[n]{x}$ as a continuous strictly increasing function on $\mathscr{R}$. (Cf. Ex. 2.)

4. If r is a rational number that is equal to the quotient of two odd positive integers, extend the domain of the function x^r, of § 1008, to the entire real number system. Discuss this function as a continuous strictly increasing function on $\mathscr{R}$. What is the nature of x^r if r has the form $r = p/q$, where p is an even positive integer and q is an odd positive integer?

5. Prove that if $\sqrt[n]{m}$, where m and n are positive integers, is rational it is integral.

6. Let n, p, and q be positive integers, and let p/q be in lowest terms. Prove that if $\sqrt[n]{p/q}$ is rational, then $\sqrt[n]{p}$ and $\sqrt[n]{q}$ are integers.

7. Prove that the subset $\mathscr{P}$ of $\mathscr{R}$ with properties as specified in Axiom IV, § 202, is unique in the following sense: Let S be a subset of $\mathscr{R}$ that is closed with respect to addition and multiplication and is such that for every real number x exactly one of the three statements $x \in S$, $x = 0$, and $-x \in S$ is true; then $S = \mathscr{P}$. Prove that a corresponding statement is true for the ordered field $\mathscr{Q}$ of rational numbers but false for the ordered field of numbers of the form $r + s\sqrt{2}$, where r and s are rational (cf. Ex. 4, § 908). Thus show that the set of numbers $r + s\sqrt{2}$ is an ordered field in two distinct ways. *Hints for $\mathscr{R}$:* If $x > 0$, then either $\sqrt{x} \in S$ or $-\sqrt{x} \in S$, whence $x \in S$; 0 is not a member of S; if $s \in S$ and if $s < 0$, then $-s > 0$ and hence $-s \in S$. *Hint for $\mathscr{Q}$:* If n is a positive integer and $-1/n \in S$, then $(-1/n)^2 = 1/n^2 \in S$ and therefore $n/n^2 = 1/n \in S$. *Hint for $r + s\sqrt{2}$:* Define S: $r + s\sqrt{2} \in S$ if and only if $r - s\sqrt{2} \in \mathscr{P}$.

8. If x and y are positive numbers not both equal to 1, and if $xy = 1$, prove that $x + y > 2$. *Hint:* Assume $x < 1$ and $y > 1$, and form the product of $1 - x$ and $y - 1$.

9. If x, y, and z are positive numbers not all equal to 1, and if $xyz = 1$, prove that $x + y + z > 3$. *Hint:* Assume that none are equal to 1 and adjust notation so that $x < 1$ and $y > 1$. Since $(1 - x)(y - 1) > 0$, $x + y > xy + 1$. By Ex. 8, since $(xy)z = 1$, $xy + z > 2$.

10. If $x_1, x_2, \cdots, x_n$ are positive numbers not all equal to 1, and if $x_1 x_2 \cdots x_n = 1$, prove that $x_1 + x_2 + \cdots + x_n > n$. *Hint:* Use Ex. 8, the ideas of Ex. 9, and induction.

11. If $x_1, x_2, \cdots, x_n$ are nonnegative numbers, their **arithmetic mean** (AM) and **geometric mean** (GM) are defined:

$$\text{AM} \equiv \frac{x_1 + x_2 + \cdots + x_n}{n}, \ \text{GM} \equiv \sqrt[n]{x_1 x_2 \cdots x_n}.$$

Prove that the arithmetic and geometric means always bear the following order relationship:

(1) $$\frac{x_1 + x_2 + \cdots + x_n}{n} \geq \sqrt[n]{x_1 x_2 \cdots x_n},$$

and that equality holds in (1) if and only if $x_1 = x_2 = \cdots = x_n$. *Hint:* In case $x_1, x_2, \cdots, x_n$ are all positive let $y_k \equiv x_k / \sqrt[n]{x_1 x_2 \cdots x_n}$, $k = 1, 2, \cdots, n$, and use Ex. 10.

12. Prove the *Schwarz-Cauchy-Buniakowski Inequality:* If $a_1, a_2, \cdots, a_n$ and $b_1 b_2, \cdots, b_n$ are arbitrary numbers, then

(2) $$\sum_{k=1}^{n} a_k b_k \leq \left(\sum_{k=1}^{n} a_k^2 \right)^{\frac{1}{2}} \cdot \left(\sum_{k=1}^{n} b_k^2 \right)^{\frac{1}{2}},$$

equality in (2) holding if and only if there exist nonnegative numbers c and d not both zero such that $ca_k = db_k$ for $k = 1, 2, \cdots, n$. (Cf. § 317, Ex. 11, § 1107.) *Hint:* If $a_1, a_2, \cdots, a_n$ are not all equal to zero, consider the quadratic function of x:

$$\sum_{k=1}^{n} (a_k x + b_k)^2 = \left(\sum_{k=1}^{n} a_k^2 \right) x^2 + 2 \left(\sum_{k=1}^{n} a_k b_k \right) x + \sum_{k=1}^{n} b_k^2,$$

and use Ex. 27, § 806.

13. Prove the *Minkowski Inequality:* If $a_1, a_2, \cdots, a_n$ and $b_1, b_2, \cdots, b_n$ are arbitrary numbers, then

(3) $$\left(\sum_{k=1}^{n} (a_k + b_k)^2 \right)^{\frac{1}{2}} \leq \left(\sum_{k=1}^{n} a_k^2 \right)^{\frac{1}{2}} + \left(\sum_{k=1}^{n} b_k^2 \right)^{\frac{1}{2}}$$

equality in (3) holding if and only if there exist nonnegative numbers c and d not both zero such that $ca_k = db_k$ for $k = 1, 2, \cdots, n$. (Cf. Ex. 12, § 1107.) *Hint:* Square both members of (3) and use Ex. 12.

14. Prove that $\sqrt[n]{n} < 2$ for every positive integer n. (Cf. Example 2, § 307; also cf. Exs. 15, 17, below.)

15. If ϵ is an arbitrary positive number, show that there exists a positive integer N such that whenever n is an integer greater than N, $\sqrt[n]{n} < 1 + \epsilon$. (Cf. Exs. 14, 17.) *Hint:* Use Ex. 22, § 308 to show that $(1 + \epsilon)^n > \frac{1}{2} n(n - 1)\epsilon^2$.

16. If $a_n \equiv \left(1 + \frac{1}{n} \right)^n$, prove that the sequence $\{a_n\}$ is strictly increasing and that for every $n > 1$, $2 < a_n < 3$. (The number e is equal to the supremum of the set $\{a_n\}$.)   *Hint:* Expand a_n by the binomial theorem (Ex. 27, § 318), and express it in the form:

$$a_n = 1 + 1 + \frac{1}{2!}\left(1 - \frac{1}{n}\right) + \frac{1}{3!}\left(1 - \frac{1}{n}\right)\left(1 - \frac{2}{n}\right) + \cdots$$

$$+ \frac{1}{n!}\left(1 - \frac{1}{n}\right)\left(1 - \frac{2}{n}\right)\cdots\left(1 - \frac{n-1}{n}\right).$$

Then $a_n < \sum_{k=0}^{n} \frac{1}{k!} \le 1 + \sum_{k=1}^{n-1} \frac{1}{2^k} = 3 - \frac{1}{2^{n-1}}.$

17. Prove that the sequence $\{\sqrt[n]{n}\}$ is strictly decreasing for $n \ge 3$.   (Cf. Exs. 14, 15.)   *Hint:* Show that the inequality $^{n+1}\sqrt{n+1} < \sqrt[n]{n}$ is equivalent to $\left(1 + \frac{1}{n}\right)^n < n$.   (Cf. Ex. 16.)

18. If n is any integer greater than 1, prove that

$$\left(\frac{n}{3}\right)^n < n! < \left(\frac{n+1}{2}\right)^n$$

Hint: A proof by mathematical induction is provided by the inequalities established in Ex. 16.

19. Use mathematical induction to prove that the nth term of the Fibonacci sequence (§ 315) is

$$a_n = \frac{(1 + \sqrt{5})^n - (1 - \sqrt{5})^n}{2^n \sqrt{5}}.$$

20. Prove that the units of an integral domain form a multiplicative commutative group.   (Cf. Exs. 30, 31, § 104.)

★21. Prove that in the integral domain Φ of § 1009, $(2 + \sqrt{5})^n$ is a unit for every integer n, and that $(2 + \sqrt{5})^m \ne (2 + \sqrt{5})^n$ if $m \ne n$.   Conclude that Φ has infinitely many distinct units.

★22. Prove that the integral domain Φ of § 1009 is dense in $\mathscr{R}$.   *Hint:* Cf. Ex. 20, § 908.

11

Exponents and Logarithms

1101. INTRODUCTION

Our objective in this chapter will be to extend the definition of a^x from all rational values of x to all real values of x. The technique for doing this will be to make use of the monotonic and continuity properties already established in Chapter 10.

In order to take care of trivial special cases we define 0^x to be 0 for all positive x, and 1^x to be 1 for all real x. (These definitions agree, for rational x, with those of §§ 1008 and 1005.)

We now turn our attention to the definition of a^x for $a > 1$ and any real number x. The details for $0 < a < 1$ and any real x are given in § 1103, but the proofs are left for the reader (Exs. 1–2, § 1107).

1102. THE FUNCTION a^x FOR $a > 1$

Definition. *If $a > 1$ and if x is real, then $a^x \equiv \sup \{a^r \mid r \in \mathcal{Q}, r < x\}$.*

Theorem I. *If x is rational this definition is consistent with that of § 1005.*

Proof. This follows from Theorem I, § 1007.

Theorem II. *If $a > 1$ and if x is real, then $a^x = \inf \{a^s \mid s \in \mathcal{Q}, s > x\}$.*

Proof. If r and s are arbitrary rational numbers such that $r < x < s$, then $a^r < a^s$ and consequently $a^s \geq \sup \{a^r \mid r \in \mathcal{Q}, r < x\} = a^x$. Therefore $a^x \leq \inf \{a^s \mid s \in \mathcal{Q}, s > x\}$. Assume that $a^x < \gamma \equiv \inf \{a^s \mid s \in \mathcal{Q}, s > x\}$, and for every natural number n let r_n and s_n be rational numbers† such that

$$(1) \qquad x - \frac{1}{2n} < r_n < x < s_n < x + \frac{1}{2n}.$$

From (1) we infer that $0 < s_n - r_n < 1/n$ and $a^{r_n} \leq a^x < \gamma \leq a^{s_n}$, and hence

$$1 < \frac{\gamma}{a^x} \leq \frac{\gamma}{a^{r_n}} \leq \frac{a^{s_n}}{a^{r_n}} = a^{s_n - r_n} < \sqrt[n]{a}$$

† Cf. the Note, § 1213.

for all natural numbers n. If $\beta \equiv (\gamma/a^x) - 1$, then $\beta > 0$, and $\sqrt[n]{a} > 1 + \beta$, or $a > (1 + \beta)^n \geq 1 + n\beta$ for all n, by the Lemma, § 1007. As in the proof of Theorem I, § 1007, the resulting inequality $n < (a - 1)/\beta$ is forbidden by the Archimedean property (§ 903), and the desired contradiction is obtained. The proof is complete.

Theorem III. *If $a > 1$ and if x is a real variable, the function a^x is strictly increasing on the set $\mathscr{R}$ of all real numbers and continuous at every real point. The values of the function a^x are all positive, and can be made arbitrarily large for positive x and less than an arbitrary positive number for negative x.*

Proof. If $x_1 < x_2$, let r and s be rational numbers such that $x_1 < r < s < x_2$. Then, by definition and Theorem II, $a^{x_1} \leq a^r < a^s \leq a^{x_2}$, and consequently $a^{x_1} < a^{x_2}$. Continuity at a point c follows from the fact that a^x is strictly increasing, from Theorem II, and from the definitions of suprema and infima (cf. Ex. 12, § 908) as follows: $a^c = \sup \{a^r \mid r \in \mathscr{2}, r < c\} \leq \sup \{a^x \mid x \in \mathscr{R}, x < c\} \leq a^c \leq \inf \{a^x \mid x \in \mathscr{R}, x > c\} \leq \inf \{a^s \mid s \in \mathscr{2}, s > c\} = a^c$. The last statement of the theorem follows immediately from the corresponding statement of Theorem I, § 1007, and the fact that a^x is strictly increasing.

NOTE. The function a^x of this section, and of the next, is called the **exponential function** with **base** a. The quantity x is called the **exponent**, and each value a^x is called a **power** of a.

1103. THE FUNCTION a^x FOR $0 < a < 1$

In this section we limit ourselves to statements only, leaving the proofs as exercises for the reader (cf. Exs. 1–2, § 1107).

Definition. *If $0 < a < 1$ and if x is real, then $a^x \equiv \inf \{a^r \mid r \in \mathscr{2}, r < x\}$.*

Theorem I. *If x is rational this definition is consistent with that of § 1005.*

Theorem II. *If $0 < a < 1$ and if x is real, then $a^x = \sup \{a^s \mid s \in \mathscr{2}, s > x\}$.*

Theorem III. *If $0 < a < 1$ and if x is a real variable, the function a^x is strictly decreasing on the set $\mathscr{R}$ of all real numbers and continuous at every real point. The values of the function a^x are all positive, and can be made arbitrarily large for negative x and less than an arbitrary positive number for positive x.*

1104. LAWS OF EXPONENTS

Theorem. *If a and b are any positive numbers and if x and y are any real numbers, then*

(i) $$a^x a^y = a^{x+y};$$

(ii) $$\frac{a^x}{a^y} = a^{x-y};$$

(iii) $$(a^x)^y = a^{xy};$$

(iv) $$(ab)^x = a^x b^x;$$

(v) $$\left(\frac{a}{b}\right)^x = \frac{a^x}{b^x}.$$

Proof. (iv) *for* $a = 1$ *or* $b = 1$: Trivial.

(iv) *for* $ab = 1$: Assume for definiteness that $a > 1$. The proposition to be established is equivalent to $a^x b^x = (1)^x = 1$, or $\dfrac{1}{a^x} = b^x = \left(\dfrac{1}{a}\right)^x$. Since $0 < \dfrac{1}{a} < 1$, the proposition takes the form that, for any x, $\gamma = \dfrac{1}{\sigma}$, where

$$\gamma \equiv \inf\left\{\left(\frac{1}{a}\right)^r \,\middle|\, r \in \mathscr{Q}, r < x\right\} \quad \text{and} \quad \sigma \equiv \sup\{a^r \,|\, r \in \mathscr{Q}, r < x\}.$$

If $\gamma < 1/\sigma$, then (by the definition of γ) there exists a rational number r such that $r < x$ and $(1/a)^r < 1/\sigma$, or $a^r > \sigma$. But this contradicts the definition of σ. If $\gamma > 1/\sigma$ so that $\sigma > 1/\gamma$, then (by the definition of σ) there exists a rational number r such that $r < x$ and $a^r > 1/\gamma$, or $(1/a)^r = 1/a^r < \gamma$. But this contradicts the definition of γ. In conclusion, we note that we have proved that *for every positive number a and real number x, $(1/a)^x = 1/a^x$, and $a^x = 1/(1/a)^x$.*

(iv) *for* $a > 1$ *and* $b > 1$: Assume that $(ab)^x \neq a^x b^x$ for some number x, and for an arbitrary natural number n let r_n and s_n be rational numbers† such that $x - \dfrac{1}{2n} < r_n < x < s_n < x + \dfrac{1}{2n}$. Then $a^{r_n} < a^x < a^{s_n}$, $b^{r_n} < b^x < b^{s_n}$, and $(ab)^{r_n} < (ab)^x < (ab)^{s_n}$. Furthermore, $(ab)^{r_n} = a^{r_n} b^{r_n} < a^x b^x < a^{s_n} b^{s_n} = (ab)^{s_n}$, and therefore $|(ab)^x - a^x b^x| < (ab)^{s_n} - (ab)^{r_n}$. Dividing by $(ab)^x$ gives:

(1) $$0 < \epsilon \equiv \left|1 - \frac{a^x b^x}{(ab)^x}\right| < \frac{(ab)^{s_n} - (ab)^{r_n}}{(ab)^x} < \frac{(ab)^{s_n} - (ab)^{r_n}}{(ab)^{r_n}}$$
$$= (ab)^{s_n - r_n} - 1 < (ab)^{1/n} - 1,$$

for every natural number n. This implies that for every n, $\sqrt[n]{ab} > 1 + \epsilon$, or $ab > (1 + \epsilon)^n \geq 1 + n\epsilon$ (Lemma, § 1007), or $n < (ab - 1)/\epsilon$. Since this contradicts the Archimedean property of the real number system (§ 903), a contradiction has been reached, and (iv) for $a > 1$ and $b > 1$ is proved.

(iv) *for* $0 < a < 1$ *and* $0 < b < 1$: By (iv) for $a > 1$ and $b > 1$ and by the last sentence of the proof of (iv) for $ab = 1$,

$$(ab)^x = \frac{1}{\left(\dfrac{1}{ab}\right)^x} = \frac{1}{\left(\dfrac{1}{a}\right)^x \left(\dfrac{1}{b}\right)^x} = a^x b^x.$$

† Cf. the Note, § 1213.

(iv) for $0 < a < 1$, $b > 1$, and $ab = c > 1$: We wish to prove that $c^x = [1/(1/a)^x]b^x$, or $(1/a)^x c^x = b^x$. But this is true by (iv) for $a > 1$ and $b > 1$ since $1/a > 1$, $c > 1$, and $b = (1/a)c$.

(iv) for $0 < a < 1$, $b > 1$, and $ab = c < 1$: We wish to prove that $1/(1/c)^x = [1/(1/a)^x]b^x$, or $(1/a)^x = (1/c)^x b^x$. But this is true by (iv) for $a > 1$ and $b > 1$ since $1/c > 1$, $b > 1$, and $1/a = (1/c)b$.

(iv) in complete generality: Because of the symmetry between a and b, the last two cases above complete the catalogue of possibilities.

(i) for $a = 1$: Trivial.

(i) for $a > 1$: Assume that there exist real numbers x and y such that $a^x a^y \neq a^{x+y}$, and for an arbitrary natural number n let r_n, s_n, u_n, and v_n be rational numbers† such that

$$(2) \quad \begin{cases} x - \dfrac{1}{4n} < r_n < x < s_n < x + \dfrac{1}{4n}, \\[2mm] y - \dfrac{1}{4n} < u_n < y < v_n < y + \dfrac{1}{4n}. \end{cases}$$

Then, from the inequalities $a^{r_n} < a^x < a^{s_n}$, $a^{u_n} < a^y < a^{v_n}$, and $r_n + u_n < x + y < s_n + v_n$ we can infer

$$(3) \qquad a^{r_n} a^{u_n} < a^x a^y < a^{s_n} a^{v_n} \quad \text{and} \quad a^{r_n + u_n} < a^{x+y} < a^{s_n + v_n}.$$

From laws of exponents for rational exponents we know that $a^{r_n} a^{u_n} = a^{r_n + u_n}$ and $a^{s_n} a^{v_n} = a^{s_n + v_n}$, and consequently $|a^x a^y - a^{x+y}| < a^{s_n + v_n} - a^{r_n + u_n}$, and hence:

$$(4) \qquad 0 < \epsilon \equiv \left| \dfrac{a^x a^y}{a^{x+y}} - 1 \right| < \dfrac{a^{s_n + v_n} - a^{r_n + u_n}}{a^{x+y}} < \dfrac{a^{s_n + v_n} - a^{r_n + u_n}}{a^{r_n + u_n}}$$

$$< a^{\frac{1}{n}} - 1.$$

From (4) we have, for every n, $a > (1 + \epsilon)^n \geq 1 + n\epsilon$ (Lemma, § 1007), in violation of the Archimedean property (§ 903).

(i) for $0 < a < 1$: By the case just proved for any base greater than 1, and by the last sentence of the proof of (iv) for $ab = 1$, we have:

$$a^x a^y = \dfrac{1}{(1/a)^x (1/a)^y} = \dfrac{1}{(1/a)^{x+y}} = a^{x+y}.$$

We note in passing that since $a^x a^{-x} = a^0 = 1$, we have proved that for any positive a and any real x, a^x and a^{-x} are reciprocals.

(ii): From (i), $a^{x-y} a^y = a^x$.

(iii) for $a = 1$: Trivial.

(iii) for $a > 1$, $xy = 0$: Trivial.

† Cf. the Note, § 1213.

(*iii*) *for* $a > 1$, $x > 0$, $y > 0$: Assume that there exist positive numbers x and y such that $(a^x)^y \neq a^{xy}$, and for an arbitrary natural number n let r_n, s_n, u_n, and v_n be *positive* rational numbers such that

(5)
$$
\begin{cases}
x - \dfrac{1}{2n} < r_n < x < s_n < x + \dfrac{1}{2n}, \\[2mm]
y - \dfrac{1}{2n} < u_n < y < v_n < y + \dfrac{1}{2n}.
\end{cases}
$$

Then $1 < a^{r_n} < a^x < a^{s_n}$, and by monotonic properties already established in §§ 1008 and 1102:

$$(a^{r_n})^{u_n} < (a^x)^{u_n} < (a^x)^y < (a^x)^{v_n} < (a^{s_n})^{v_n},$$

so that (by (*iii*) of § 1006): $a^{r_n u_n} < (a^x)^y < a^{s_n v_n}$. Furthermore, since $r_n u_n < xy < s_n v_n$, $a^{r_n u_n} < a^{xy} < a^{s_n v_n}$, and hence $|(a^x)^y - a^{xy}| < a^{s_n v_n} - a^{r_n u_n}$. Therefore:

(6)
$$0 < \epsilon \equiv \left| \frac{(a^x)^y - a^{xy}}{a^{xy}} \right| < \frac{a^{s_n v_n} - a^{r_n u_n}}{a^{xy}} < \frac{a^{s_n v_n} - a^{r_n u_n}}{a^{r_n u_n}}$$

$$< a^{\left(r_n + \frac{1}{n}\right)\left(u_n + \frac{1}{n}\right) - r_n u_n} - 1$$

$$\leqq a^{(r_n + u_n + 1)/n} - 1 < \sqrt[n]{a^{x+y+1}} - 1.$$

But this means that $a^{x+y+1} > (1 + \epsilon)^n \geqq 1 + n\epsilon$ for every natural number n, again in violation of the Archimedean property.

(*iii*) *for* $a > 1$, $x < 0$, *and* $y < 0$: Using the remarks at the end of the proof of (*i*) and at the end of the proof of (*iv*) *for* $ab = 1$, and (*iii*) *for* $a > 1$, $x > 0$, *and* $y > 0$, we have: $(a^x)^y = 1/[1/a^{-x}]^{-y} = 1/[1/(a^{-x})^{-y}] = 1/[1/a^{xy}] = a^{xy}$.

(*iii*) *for* $a > 1$, $x < 0$, *and* $y > 0$: As in the preceding proof, $(a^x)^y = [1/a^{-x}]^y = 1/(a^{-x})^y = 1/a^{-xy} = a^{xy}$.

(*iii*) *for* $a > 1$, $x > 0$, *and* $y < 0$: As above, $(a^x)^y = 1/(a^x)^{-y} = 1/a^{-xy} = a^{xy}$.

(*iii*) *for* $a \geqq 1$: Now completely proved.

(*iii*) *for* $0 < a < 1$: By the note at the end of the proof of (*iv*) *for* $ab = 1$, and (*iii*) *for* $a > 1$, we have: $(a^x)^y = [1/(1/a)^x]^y = 1/[(1/a)^x]^y = 1/(1/a)^{xy} = a^{xy}$.

(*v*): From (*iv*), $b^x(a/b)^x = a^x$.

1105. THE FUNCTION x^c

As in § 1008 we state and prove a theorem concerning the function x^c for fixed $c > 0$ and variable $x \geqq 0$, and state without proof a corresponding theorem for fixed $c < 0$ and variable $x > 0$ (cf. Ex. 3, § 1107).

Theorem I. *If c is a positive number and if x is a variable restricted to nonnegative values, the function x^c is strictly increasing on the interval $I = [0, +\infty)$ and continuous at every point x of I. The values of x^c are nonnegative and unbounded, and include $0^c = 0$ and $1^c = 1$.*

Proof. If $0 < y$, then y^c is positive (and hence $0^c < y^c$), by definition of y^1, or by Theorem III, § 1102, or by Theorem III, § 1103. If $0 < x < y$, the inequality $x^c < y^c$ is equivalent, by (v), § 1104 to $(y/x)^c > 1$; with $a \equiv y/x$, the inequality $a^c > 1 = a^0$ follows from Theorem III, § 1102, and the inequality $c > 0$. Continuity of x^c at $x = 0$ means that $0 = \inf \{x^c \mid x > 0\}$. Clearly, $0 \leq \inf \{x^c \mid x > 0\}$. Assume that $0 < \gamma \equiv \inf \{x^c \mid x > 0\}$. Then for every $x > 0$, $x^c \geq \gamma$, or (by the monotonicity proved above) $x = (x^c)^{1/c} \geq \gamma^{(1/c)} > 0$. But this is manifestly false for $x = \frac{1}{2}\gamma^{(1/c)}$. (Contradiction.) To establish continuity of x^c at an arbitrary positive number a, we wish to prove:

(1) $\sigma \equiv \sup \{x^c \mid 0 \leq x < a\} = a^c = \gamma \equiv \inf \{x^c \mid x > a\}$.

Obviously, by monotonicity, $0 < \sigma \leq a^c \leq \gamma$. If $\sigma < a^c$, then $0 \leq x < a$ implies $x^c \leq \sigma < a^c$, or (by the monotonicity proved above) $x \leq \sigma^{1/c} < a$. (Contradiction.) If $a^c < \gamma$, then $x > a$ implies $x^c \geq \gamma > a^c$, or $x \geq \gamma^{1/c} > a$. (Contradiction.) Therefore (1) is proved. The last statement of Theorem I follows immediately from the corresponding statement of Theorem I, § 1008, and the fact that x^c is strictly increasing.

Theorem II. *If c is a negative number and if x is a variable restricted to positive values, the function x^c is strictly decreasing on the interval $I = (0, +\infty)$ and continuous at every point x of I. The values of x^c are positive, and can be made arbitrarily large for small positive x and less than an arbitrary positive number for large positive x. The values of x^c include $1^c = 1$.*

Proof. The proof is left as an exercise for the reader (Ex. 3, § 1107).

NOTE. The function x^c is called a **power function** of x.

1106. LOGARITHMS

Let a be a fixed positive number not equal to 1, and let x be any positive number. The problem before us in this section is to show first that there exists one and only one number y such that $a^y = x$, and second to investigate the properties of the function of x thus determined and written:

(1) $y = \log_a x$ *if and only if $a^y = x$.*

The *function* $\log_a x$ is called the **logarithmic function** with **base** a, and the *number* $\log_a x$, for a given x, is called the **logarithm** of x to the base a.

We shall limit explicit discussion to the case $a > 1$, leaving the case $0 < a < 1$ to the reader as an exercise (Ex. 5, § 1107).

Theorem I. *If $a > 0$, $a \neq 1$, and $x > 0$, then there exists one and only one number y such that $a^y = x$.*

Proof for $a > 1$. By Theorem III, § 1102, there exist numbers b and c such that $a^b < x < a^c$; and therefore the set

$$A \equiv \{u \mid u \in \mathscr{R}, a^u < x\}$$

is nonempty ($b \in A$) and bounded above (by c). If $y \equiv \sup A$, we shall prove that $a^y = x$ by showing that each of the two inequalities $a^y < x$ and $a^y > x$ leads to a contradiction. We first assume $a^y < x$. Then, by Theorem III, § 1102, since $a^y = \inf \{a^v \mid v \in \mathscr{R}, v > y\}$, there must exist a number v such that $v > y$ and $a^v < x$. This last inequality implies that v is a member of A, which is impossible if $v > y$ since y is an upper bound of A. (Contradiction.) We now assume that $a^y > x$ and use the fact (cf. Theorem III, § 1102) that $a^y = \sup \{a^w \mid w \in \mathscr{R}, w < y\}$. There exists, then, a number w such that $w < y$ and $a^w > x$. Since the exponential function with base greater than 1 is an increasing function, it follows that w is an upper bound of the set A (if $u > w$, then $a^u > a^w > x$). This is impossible since $w < y$ and y is the *least* upper bound of A. (Contradiction.) Finally, uniqueness follows from the strictly increasing nature of the exponential function with base greater than 1: If $a^{y_1} = a^{y_2} = x$, and if $y_1 < y_2$, then $a^{y_1} < a^{y_2}$. (Contradiction.)

Corollary I. *If $a > 0$ and $a \neq 1$ and if b is any positive number, then:*

$$(2) \qquad \log_a (a^b) = a^{\log_a b} = b.$$

Proof. In the first place, by definition $\log_a (a^b)$ is the unique number y such that $a^y = a^b$, and is therefore equal to b. In the second place, again by definition, since $y = \log_a b$ if and only if $a^y = b$, it follows that $a^{\log_a b} = b$.

Corollary II. *If $a > 0$ and $a \neq 1$,*

$$(3) \qquad \log_a a = 1, \quad \log_a 1 = 0.$$

Proof. This follows from (1) since $a^1 = a$ and $a^0 = 1$.

The following theorem is stated for simplicity only for the logarithmic function with base a greater than 1. For the case $0 < a < 1$ see Exercise 7, § 1107.

Theorem II. *If $a > 1$ the logarithmic function $\log_a x$ is strictly increasing on the set $\mathscr{P}$ of positive numbers and is continuous at every positive point. Its values are arbitrarily large for large x and are negative with arbitrarily large absolute value for small positive x.*

Proof. The strict increase follows directly from the corresponding behavior for the exponential function: If $y_1 = \log_a x_1$, $y_2 = \log_a x_2$, and $x_1 < x_2$, then $y_1 < y_2$ since from $x_1 = a^{y_1}$ and $x_2 = a^{y_2}$ the inequality $y_1 \geqq y_2$ would imply $x_1 \geqq x_2$.

To prove continuity at the point x we wish to establish the inequalities, where all numbers involved are members of $\mathscr{R}$:

(4) $\log_a x = \sup \{\log_a u \mid 0 < u < x\}$,

(5) $\log_a x = \inf \{\log_a v \mid v > x\}$.

Since $0 < u < x < v$ implies $\log_a u < \log_a x < \log_a v$,

(6) $\sup \{\log_a u \mid 0 < u < x\} \leq \log_a x \leq \inf \{\log_a v \mid v > x\}$.

Assume first that $\log_a x > \sigma$, where $\sigma \equiv \sup \{\log_a u \mid 0 < u < x\}$. The statement that $\sigma \geq \log_a u$ for every positive number u less than x is equivalent to the statement that $a^\sigma \geq u$ for every positive number u less than x. But this is ridiculous since $a^\sigma < x$ (choose u to be any number between a^σ and x). (Contradiction.)   Assume now that $\log_a x < \gamma$, where $\gamma \equiv \inf \{\log_a v \mid v > x\}$. The statement that $\gamma \leq \log_a v$ for every number v greater than x is equivalent to the statement that $a^\gamma \leq v$ for every positive number v greater than x. But this is ridiculous since $a^\gamma > x$ (choose v to be any number between a^γ and x). (Contradiction.) Therefore equalities hold in (6), and continuity is established.

The statements of the last sentence of the theorem follow from the strictly increasing nature of the logarithmic function and the properties of the exponential function stated in Theorem III, § 1102.

Theorem III. *The following laws of logarithms hold, where* a, b, x, *and* y *are positive numbers and* a *and* b *are different from* 1:

(i) $\log_a (xy) = \log_a x + \log_a y$.

(ii) $\log_a \left(\dfrac{x}{y}\right) = \log_a x - \log_a y$.

(iii) $\log_a (x^y) = y \log_a x$.

(iv) $\log_a x = \log_a b \log_b x$.

(v) $\log_a x = \dfrac{\log_b x}{\log_b a}$.

(vi) $\log_a b = \dfrac{1}{\log_b a}$.

Proof.   These properties follow from corresponding properties of exponents given in § 1104, as follows: (i): Let $u \equiv \log_a x$, $v \equiv \log_a y$, and $w \equiv \log_a (xy)$. Then $xy = a^w = a^u a^v = a^{u+v}$, and $w = u + v$. (ii): From (i), $\log_a x = \log_a [y(x/y)] = \log_a y + \log_a (x/y)$. (iii): Let $u \equiv \log_a x$ and $v \equiv \log_a (x^y)$. Then $x^y = a^v = (a^u)^y = a^{uy}$, and $v = uy$. (iv): Let $u \equiv \log_a x$, $v \equiv \log_a b$, and $w \equiv \log_b x$. Then $a^u = x$, $a^v = b$, and $b^w = x$, so that $x = a^u = (a^v)^w = a^{vw}$, and $u = vw$. (v): This is equivalent to (iv) with a and b interchanged. (vi): This is a special case of (v), with $x = b$.

1107. EXERCISES

In these exercises all numbers are assumed to be real.

1. Prove Theorem II, § 1103.

2. Prove Theorem III, § 1103.

3. Prove Theorem II, § 1105.

4. In the definition of $\log_a x$, why is the case $a = 1$ excluded? What about $\log_1 1$?

5. Prove Theorem I, § 1106, for the case $0 < a < 1$.

6. If $a > 0$, $a \neq 1$, and $x > 0$, prove that $\log_a x = -\log_{\frac{1}{a}} x$.

7. State a theorem regarding the monotonic behavior and the continuity of the logarithmic function $\log_a x$ corresponding to Theorem II, § 1106, for the case $0 < a < 1$.   *Hint:* Cf. Ex. 6.

8. If $a > 1$ and $x > 1$, prove that $(\log_a x)^{\log_a x} = x^{\log_a(\log_a x)}$.

9. If a is any positive constant different from 1, prove that the exponential function a^x is **convex**: If x_1 and x_2 are any real numbers, and if α and β are positive numbers whose sum is 1, then

(1) $$a^{\alpha x_1 + \beta x_2} \leq \alpha a^{x_1} + \beta a^{x_2},$$

equality in (1) holding if and only if $x_1 = x_2$. (Cf. Ex. 17, § 204, and the references on inequalities given in the footnote of § 202.   *Hints:* Let P be the proposition just stated and let P' be the same proposition with the omission of the last phrase involving equality.   (*i*) Prove that P in general is equivalent to P with the base a restricted to values greater than 1 (if (1) holds for $a > 1$ and if $0 < d < 1$, let $a = 1/d$, and adjust notation).   (*ii*) Assuming $x_1 < x_2$ and letting $b \equiv a^{x_2 - x_1}$, prove that P is equivalent to the inequality

(2) $$b^\beta < 1 + (b - 1)\beta, \quad \text{for} \quad b > 1, 0 < \beta < 1.$$

(*iii*) By letting $x = 1/\beta$ and $c \equiv 1 + (b - 1)/x$, prove that P is equivalent to the inequality

(3) $$c^x > 1 + (c - 1)x, \quad \text{for} \quad c > 1, x > 1.$$

(*iv*) Use the arithmetic mean-geometric mean inequality of Ex. 11, § 1010 for $m + n$ positive numbers of which m are equal to 1 and n are equal to $1 + (m + n)(c - 1)/n$ to prove (3) for all rational $x > 1$.   (*v*) Establish the inequality

(4) $$c^x \geq 1 + (c - 1)x, \quad \text{for} \quad c > 1, x > 1$$

by assuming $c^x < 1 + (c - 1)x$ for some $c > 1$ and $x > 1$, and choosing a rational number $r > x$ such that $c^r < 1 + (c - 1)x$ (cf. Theorem II, § 1102).   (*vi*) Prove P' by showing that P' is equivalent to (4).   (*vii*) Establish (3) by assuming the existence of $c > 1$ and $x > 1$ such that $c^x = 1 + (c - 1)x$ and let r be a rational number between 1 and x, with $r = \alpha + \beta x$, $\alpha > 0$, $\beta > 0$, $\alpha + \beta = 1$. Use P' in the form $c^r \leq \alpha c + \beta c^x = 1 + (c - 1)r$, in contradiction to (3) for rational exponents.

10. If $x \geq 0$, $y \geq 0$, $0 < \alpha < 1$, $0 < \beta < 1$, and $\alpha + \beta = 1$, prove that $x^\alpha y^\alpha \leq \alpha x + \beta y$ and that equality holds if and only if $x = y$.   *Hint:* If x and y are any two distinct positive numbers, let a be any positive number different from 1, and let $u \equiv \log_a x$ and $v \equiv \log_a v$. Use the inequality $a^{\alpha u + \beta v} < \alpha a^u + \beta a^v$ from (1), Ex. 9.

11. Prove the **Hölder inequality** for finite sums, where $p > 1$, $q > 1$, $\dfrac{1}{p} + \dfrac{1}{q} = 1$, and $a_1, a_2, \cdots, a_n$ and $b_1, b_2, \cdots, b_n$ are arbitrary:

$$(5) \qquad \sum_{k=1}^{n} a_k b_k \leq \left(\sum_{k=1}^{n} |a_k|^p \right)^{\frac{1}{p}} \left(\sum_{k=1}^{n} |b_k|^q \right)^{\frac{1}{q}}$$

equality in (5) holding if and only if $a_k b_k \geq 0$ for $k = 1, 2, \cdots, n$ and there exist nonnegative numbers c and d not both zero such that $c|a_k|^p = d|b_k|^q$ for $k = 1, 2, \cdots, n$. (For the special case $p = q = 2$, cf. Ex. 12, § 1010.) *Hint:* In case the right-hand member of (5) is positive, use the inequality of Ex. 10 for each $k = 1, 2, \cdots, n$, with $\alpha = 1/p$, $\beta = 1/q$,

$$x = \frac{|a_k|^p}{\displaystyle\sum_{k=1}^{n} |a_k|^p}, \qquad y = \frac{|b_k|^q}{\displaystyle\sum_{k=1}^{n} |b_k|^q},$$

and add.

12. Prove the **Minkowski inequality** for finite sums, where $p \geq 1$ and $a_1, a_2, \cdots, a_n$ and $b_1, b_2, \cdots, b_n$ are arbitrary:

$$(6) \qquad \left(\sum_{k=1}^{n} |a_k + b_k|^p \right)^{\frac{1}{p}} \leq \left(\sum_{k=1}^{n} |a_k|^p \right)^{\frac{1}{p}} + \left(\sum_{k=1}^{n} |b_k|^p \right)^{\frac{1}{p}}.$$

If $p = 1$ equality in (6) holds if and only if $a_k b_k \geq 0$ for $k = 1, 2, \cdots, n$. If $p > 1$, equality in (6) holds if and only if there exist nonnegative numbers c and d not both zero such that $c a_k = d b_k$ for $k = 1, 2, \cdots, n$. (For the special case $p = 2$, cf. Ex. 13, § 1010.) *Hint:* For $p > 1$, define $q \equiv p/(p-1)$, $c_k \equiv |a_k|$, $d_k \equiv |b_k|$, $k = 1, 2, \cdots, n$. Then apply Hölder's inequality (Ex. 11) as follows:

$$\sum (c_k + d_k)^p = \sum c_k (c_k + d_k)^{p-1} + \sum d_k (c_k + d_k)^{p-1}$$

$$\leq \left(\sum c_k{}^p \right)^{\frac{1}{p}} \left(\sum \left(c_k + d_k \right)^p \right)^{\frac{1}{q}} + \left(\sum d_k{}^p \right)^{\frac{1}{p}} \left(\sum \left(c_k + d_k \right)^p \right)^{\frac{1}{q}}.$$

13. If $a_1, a_2, \cdots, a_n$ are any n numbers, their **root-mean-square** is defined by the formula $\sqrt{\dfrac{1}{n} \displaystyle\sum_{k=1}^{n} a_k{}^2}$. Prove that the arithmetic mean of the n numbers $a_1, a_2, \cdots, a_n$ is less than or equal to their root-mean-square (cf. Ex. 11, § 1010):

$$(7) \qquad \frac{1}{n} \sum_{k=1}^{n} a_k \leq \sqrt{\frac{1}{n} \sum_{k=1}^{n} a_k{}^2},$$

equality in (7) holding if and only if the a_k's are all equal and nonnegative. More generally, establish the inequality

$$(8) \qquad \frac{1}{n} \sum_{k=1}^{n} a_k \leq \left(\frac{1}{n} \sum_{k=1}^{n} |a_k|^p \right)^{\frac{1}{p}},$$

equality in (8) holding if and only if the a_k's are all equal and nonnegative. *Hint:* Use the Schwarz-Cauchy-Buniakowski inequality of Ex. 12, § 1010, and the Hölder inequality of Ex. 11, above, with the b_k's all equal to 1.

12

Decimal Expansions

++

1201. DECIMAL REPRESENTATION OF NATURAL NUMBERS

The real numbers are commonly represented in terms of powers of the natural number ten, since man did his first counting and reckoning with the aid of his ten fingers. From the mathematical viewpoint any natural number greater than 1 could be used as well as ten for this **base** or **radix** of the number system. With bases of two, three, four, five, ten, and twelve the corresponding system of representing numbers is called **binary** or **dyadic, ternary, quaternary, quintic, decimal**, and **duodecimal**, respectively. Each such representation of a number is called a **numeral**. Our purpose in the present section is to develop a procedure for obtaining a decimal representation for an arbitrary given natural number. Although the text discussion is focused on the base ten, the method is general. Other bases will be brought into the examples and exercises.

When we write a number like 35,027 in the decimal system we are using a compact form for

$$3 \cdot 10^4 + 5 \cdot 10^3 + 0 \cdot 10^2 + 2 \cdot 10^1 + 7 \cdot 10^0.$$

The integers 3, 5, 0, 2, and 7 are called the *digits* of the given number, 3 being the *ten-thousands digit*, 5 the *thousands digit*, 0 the *hundreds digit*, 2 the *tens digit* and 7 the *units digit*. The number 35,027 is thereby represented as a sum of terms each of which is a nonnegative integral digit less than 10 multiplied by a power of 10 with nonnegative integral exponent. That every natural number can be so represented—and uniquely so represented—is the substance of the next theorem. Our work will be simplified if we first introduce a definition and an item of notation.

Definition. *Let a be a nonnegative integer, and let q and r be the unique nonnegative integers such that*

(1) $r < 10$

and

(2) $a = 10q + r.$

(Cf. Theorem II, § 403.) *Then r is called the* **units digit** *of the decimal representation of a.*

For notational convenience in this chapter we recall the greatest integer function defined in the Note § 903:

Notation. *If x is any real number, denote by [x] the unique integer* $n = [x]$ such that

(3) $n \leq x < n + 1,$ or $[x] \leq x < [x] + 1,$ or $x - 1 < [x] \leq x,$

as guaranteed by Theorem II, § 903. As a function of x, $[x]$ is called the **greatest integer function** or the **bracket function**, and prescribes the greatest integer less than or equal to x. If x is nonnegative, then so is $[x]$. *In this chapter, square brackets [ ] will be used only for the bracket function.*

Theorem. *Every natural number a has a unique decimal representation in the form*

(4) $$a = d_{-m}10^m + d_{-m+1}10^{m-1} + \cdots + d_{-k}10^k + \cdots$$
$$+ d_{-1}10^1 + d_0 10^0,$$

where $d_{-k} = 0, 1, 2, 3, 4, 5, 6, 7, 8,$ *or* 9 *for* $0 \leq k \leq m,$ *and where* d_{-m} *is nonzero. Each* **coefficient** *or* **digit** d_{-k} *in* (4) *is equal to the units digit of the nonnegative integer* $[10^{-k}a],$ $k = 0, 1, \cdots, m.$

Proof. Existence: Let $P(n)$ be the statement that a representation of the form (4) is possible for every natural number a less than 10^n. Clearly $P(1)$ is true. Assuming that $P(n)$ is true for a particular natural number n, let a be a natural number less than 10^{n+1} and write a, by means of the Fundamental Theorem of Euclid (§ 403): $a = 10q + r$, where q and r are nonnegative integers and $r < 10$. It follows that $q < 10^n$ and hence, by the induction assumption, that q can be written in the form (4). Consequently upon expansion of $a = 10(d_{-m}10^m + \cdots + d_{-1}10 + d_0) + r$, we obtain the desired form. Since every natural number a is less than 10^n for *some* n (for example, for $n = a$, as shown in Example 2, § 307), it follows from the Fundamental Theorem of Mathematical Induction that $P(n)$ is true for all natural numbers n, and the representation (4) exists for all natural numbers a.

Uniqueness: If a is a natural number given by (4), then, since $d_{-m} \geq 1$, $a \geq d_{-m}10^m \geq 10^m$, and since $d_{-k} \leq 9$ for $0 \leq k \leq m$, $a \leq 9 \cdot 10^m + 9 \cdot 10^{m-1} + \cdots + 9$ and, by Theorem II, § 314, for the sum of a geometric progression, this last sum is equal to:

$$9 \cdot 10^m \frac{1 - (\tfrac{1}{10})^{m+1}}{1 - \tfrac{1}{10}} < 9 \cdot 10^m \frac{1}{\tfrac{9}{10}} = 10^{m+1}.$$

In other words, a satisfies the inequalities $10^m \leq a < 10^{m+1}$, and m is uniquely determined by a. Furthermore, for any integer k such that $0 \leq k \leq m$:

(5) $10^{-k}a = d_{-m}10^{m-k} + \cdots + d_{-k-1}10 + d_{-k} + \dfrac{d_{-k+1}}{10} + \cdots + \dfrac{d_0}{10^k}$.

Since

$$\dfrac{d_{-k+1}}{10} + \cdots + \dfrac{d_0}{10^k} \leq \dfrac{9}{10} + \cdots + \dfrac{9}{10^k} = \dfrac{9}{10}\left(1 + \dfrac{1}{10} + \cdots + \left(\dfrac{1}{10}\right)^{k-1}\right),$$

we have from Theorem II, § 314:

$$\dfrac{d_{-k+1}}{10} + \cdots + \dfrac{d_0}{10^k} \leq \dfrac{9}{10}\dfrac{1 - (\frac{1}{10})^k}{1 - \frac{1}{10}} < \dfrac{9}{10}\dfrac{1}{\frac{9}{10}} = 1,$$

and consequently $[10^{-k}a] = d_{-m}10^{m-k} + \cdots + d_{-k-1}10 + d_{-k}$. From this it follows that the units digit of $[10^{-k}a]$ is d_{-k}. Since m is uniquely determined by a, and since for each $k = 0, 1, \cdots, m$ the digit d_{-k} is also uniquely determined by a, the proof is complete.

Example 1. Express the decimal numeral 301 in terms of powers of 2, and hence write it as a numeral in the binary system (in which the only digits are 0 and 1).

First solution. In decimal notation, since $256 < 301 < 512$, we write $301 = 256 + 45$. Since $32 < 45 < 64$, we write $45 = 32 + 13$. Continuing in this fashion and combining results, we have

$$301 = 256 + 32 + 8 + 4 + 1$$
$$= 2^8 + 2^5 + 2^3 + 2^2 + 2^0.$$

The binary numeral desired is therefore 100,101,101.

Second solution. The units digit is the remainder after division by 2, and is equal to 1, since (in decimal notation)

$$301 = 150 \cdot 2 + 1.$$

The twos digit is the remainder after division of the quotient 150 by 2 (in this case the word quotient is used in the sense defined in Note 1, § 403), and this remainder is 0: $150 = 75 \cdot 2 + 0$. This process can be continued, dividing 75 by 2, etc. We systematize the entire sequence of divisions in a descending pattern, with remainders recorded at the right, thus:

	Remainders
2)301	
2)150	1
2) 75	0
2) 37	1
2) 18	1
2) 9	0
2) 4	1
2) 2	0
2) 1	0
0	1

Since the remainders are the digits in the binary system, with the units digit at the top, the binary numeral is obtained in agreement with the first solution: 100,101,101.

Example 2. Find the quinary numeral for the decimal numeral 3529, and check.

Solution. We follow the method of the second solution of Example 1:

	Remainders
5)3529	
5) 705	4
5) 141	0
5) 28	1
5) 5	3
5) 1	0
0	1

Therefore the desired quinary numeral is 103,104. That this is correct can be seen by expressing this result in powers of 5 in the decimal notation: $3125 + 3 \cdot 125 + 25 + 4 = 3529$.

1202. ADDITION, SUBTRACTION, MULTIPLICATION, AND DIVISION

The standard procedures for adding, subtracting, multiplying, and dividing natural numbers when they are represented as decimal numerals are familiar to school children. It is possible, however, that the "reasons why" are not always thoroughly appreciated, and a second look may be worthwhile. Since the processes of arithmetic expressed in terms of the base ten have become so mechanically ingrained that they are often carried out automatically without much actual thought being needed, it is instructive to force ourselves to think about what we are doing by means of a change of base. In many of the examples and exercises that follow the base will be something other than ten.

Addition and subtraction are both fairly simple, the most complicated ideas being those of "carrying" and "borrowing." We shall limit our discussion to two examples, one in the quinary system and one in the binary system.

Example 1. Add the three numbers whose decimal numerals are 38, 96, and 68 by changing them to base five.

Solution. The three numbers expressed as quinary numerals are 123, 341, and 233. Notice that in the addition given below, 1 is carried from the first column to the second, 2 from the second column to the third, and 1 from the third column to a new fourth column.

$$
\begin{array}{r}
123 \\
341 \\
233 \\
\hline
1302
\end{array}
$$

The sum, 1302 in the quinary system, is represented as $125 + 75 + 2 = 202$ in decimal notation.

Example 2. Use the binary system to perform the following subtraction, where the given numerals are decimal: $4181 - 2027 = 2154$.

Solution.   1000001010101
                   11111101011
                   _____
                   100001101010.

To check, we express the answer in decimal notation: $2^{11} + 2^6 + 2^5 + 2^3 + 2 = 2048 + 64 + 32 + 8 + 2 = 2154$.

Multiplication brings in the distributive law, as illustrated in Example 3. If the base b is different from ten, any multiplication problem reduces to repeated use of a basic multiplication table of the numbers $0, 1, 2, \cdots, b - 1$. In usual practice this is simple enough to be done in one's head, without special tables or paper work. (For example, if b is seven, then $3 \cdot 3 = 12$ and $2 \cdot 5 = 13$.)†

Example 3. Use the laws of real numbers to justify the following formal procedure for multiplying 68 by 27 (in decimal notation). Then perform the same multiplication in the ternary and quinary systems.

$$\begin{array}{r} 68 \\ 27 \\ \hline 476 \\ 136 \\ \hline 1836. \end{array}$$

Solution. The product can be written and expanded:

$$\begin{aligned} (6 \cdot 10 + 8)(2 \cdot 10 + 7) &= (6 \cdot 10 + 8) \cdot 7 + (6 \cdot 10 + 8) \cdot 2 \cdot 10 \\ &= \{42 \cdot 10 + 56\} + \{12 \cdot 100 + 16 \cdot 10\} \\ &= \{4 \cdot 10^2 + 7 \cdot 10 + 6\} + \{1 \cdot 10^3 + 3 \cdot 10^2 + 6 \cdot 10\} \\ &= 1 \cdot 10^3 + 7 \cdot 10^2 + 13 \cdot 10 + 6 \\ &= 1 \cdot 10^3 + 8 \cdot 10^2 + 3 \cdot 10 + 6. \end{aligned}$$

In the ternary and quinary systems the given multiplication takes the following two forms:

$$\begin{array}{r} 2112 \\ 1000 \\ \hline 2112000 \\ \textit{Ternary} \end{array} \qquad \begin{array}{r} 233 \\ 102 \\ \hline 1021 \\ 2330 \\ \hline 24321 \\ \textit{Quinary} \end{array}$$

† A subscript is sometimes a convenient means of indicating a base other than ten. For example, the representations obtained in Examples 1 and 2, § 1201, can be written $301 = (100, 101, 101)_2$ and $3529 = (103, 104)_5$, respectively, while the products just mentioned with base seven take the form $3 \cdot 3 = (3)_7 \cdot (3)_7 = (12)_7$ and $2 \cdot 5 = (2)_7 \cdot (5)_7 = (13)_7$.

	1	2	3	4	5	6	7	8	9	t	e
1	2	3	4	5	6	7	8	9	t	e	10
2	3	4	5	6	7	8	9	t	e	10	11
3	4	5	6	7	8	9	t	e	10	11	12
4	5	6	7	8	9	t	e	10	11	12	13
5	6	7	8	9	t	e	10	11	12	13	14
6	7	8	9	t	e	10	11	12	13	14	15
7	8	9	t	e	10	11	12	13	14	15	16
8	9	t	e	10	11	12	13	14	15	16	17
9	t	e	10	11	12	13	14	15	16	17	18
t	e	10	11	12	13	14	15	16	17	18	19
e	10	11	12	13	14	15	16	17	18	19	1t

Addition

	2	3	4	5	6	7	8	9	t	e
2	4	6	8	t	10	12	14	16	18	1t
3	6	9	10	13	16	19	20	23	26	29
4	8	10	14	18	20	24	28	30	34	38
5	t	13	18	21	26	2e	34	39	42	47
6	10	16	20	26	30	36	40	46	50	56
7	12	19	24	2e	36	41	48	53	5t	65
8	14	20	28	34	40	48	54	60	68	74
9	16	23	30	39	46	53	60	69	76	83
t	18	26	34	42	50	5t	68	76	84	92
e	1t	29	38	47	56	65	74	83	92	t1

Multiplication

Each result can be checked by reconverting to the decimal system. The ternary product is equal to $1458 + 243 + 81 + 54 = 1836$, and the quinary product is equal to $1250 + 500 + 75 + 10 + 1 = 1836$.

The standard procedure or algorithm of dividing one natural number by another is again justified by expressing the various constituent parts of the total array in terms of powers of the base.

Example 4. Use the laws of real numbers to justify the following formal procedure for dividing 1955 by 23 (in decimal notation). Then perform the same division in the systems with base nine and base seven.

$$
\begin{array}{r}
85 \\
23\overline{)1955} \\
184 \\
\hline
115 \\
115.
\end{array}
$$

Solution. $\quad 1955 = 184 \cdot 10 + 115$
$\qquad\qquad = 23 \cdot 80 + 23 \cdot 5$
$\qquad\qquad = 23(8 \cdot 10 + 5) = 23 \cdot 85.$

With base nine and base seven the divisions become:

$$
\begin{array}{r}
104 \\
25\overline{)2612} \\
25 \\
\hline
112 \\
112 \\
\hline
\end{array}
\qquad\qquad
\begin{array}{r}
151 \\
32\overline{)5462} \\
32 \\
\hline
226 \\
223 \\
\hline
32 \\
32 \\
\hline
\end{array}
$$

$$\textit{Base nine} \qquad\qquad \textit{Base seven}$$

Example 5. Write out the addition and multiplication tables for the duodecimal system, using t and e for ten and eleven, respectively. Then perform the multiplication $285 \times 562 = 160{,}170$ (base ten) by converting to the duodecimal system, and check.

Solution. The addition and multiplication tables (with 0 omitted from the addition table and with 0 and 1 omitted from the multiplication table) are on the opposite page.

Conversion of the factors in the product to the duodecimal system takes the form, in decimal notation:

	Remainders		*Remainders*
12)285		12)562	
12)23	9	12)46	10
12)1	11	12)3	10
0	1,	0	3,

so that the problem in duodecimal notation is to perform the multiplication
$1e9 \times 3tt$:

$$
\begin{array}{r}
1e9 \\
3tt \\
\hline
1796 \\
1796 \\
5e3 \\
\hline
78{,}836.
\end{array}
$$

To check the result we convert it back to the decimal system: $7 \cdot 12^4 + 8 \cdot 12^3 + 8 \cdot 12^2 + 3 \cdot 12 + 6 = 7 \cdot 20736 + 8 \cdot 1728 + 8 \cdot 144 + 3 \cdot 12 + 6 = 145{,}152 + 13{,}824 + 1152 + 36 + 6 = 160{,}170$.

1203. TERMINATING DECIMALS

The rational numbers of the form $p/10^q$, where p is an integer and q is a nonnegative integer are called **decimal rationals, decimal-rational numbers, or decimal fractions**. If p is a positive integer less than 10^q, then $p/10^q$ is a decimal fraction between 0 and 1. In notational systems other than decimal the terminology is similar. For example, in the binary system one speaks of binary rationals, binary-rational numbers, and binary fractions.

Since computational and numerical work calls for approximations by finite decimal expansions (and by similar expansions with bases other than ten) the following theorem is important (cf. § 904 for the definition of denseness):

Theorem I. *The decimal rationals are dense in $\mathscr{R}$, and similarly for bases other than ten.*

Proof. Let x and y be any real numbers such that $x < y$, let q be a positive integer such that $\dfrac{1}{q} < y - x$. and let p be an integer such that

$$
x < \frac{p}{10^q} \leqq x + \frac{1}{10^q} < y.
$$

The remaining ideas of the proof of Theorem I, § 904, including the existence of both q and p, apply to the present proof since $1/10^q < 1/q$.

Since decimal expansions of negative numbers can be immediately obtained from decimal expansions of positive numbers, and since any positive number that is not an integer can be expressed as the sum of a nonnegative integer and a positive number less than 1, we shall concentrate principally on the problem of obtaining decimal representations for numbers a satisfying the inequalities $0 < a < 1$. In the present section we consider the decimal rationals in the interval $(0, 1)$.

Notation.   *The symbol* $.d_1d_2 \cdots d_m$ *denotes the sum*

(1)
$$a = \frac{d_1}{10} + \frac{d_2}{10^2} + \cdots + \frac{d_k}{10^k} + \cdots + \frac{d_m}{10^m},$$

where the **coefficient** or **digit** d_k is a nonnegative integer less than 10, for $k = 1, 2, \cdots, m$, and where $d_m \neq 0$.

Any sum of the form (1) or, more generally, any nonnegative integer or a nonnegative integer plus a sum of the form (1) is called a **terminating decimal**.

Theorem II.   *Every terminating decimal of the form* (1) *is a decimal fraction less than* 1. *Conversely, every decimal fraction a between* 0 *and* 1 *can be uniquely represented as a terminating decimal in the form* (1). *The digit d_k in* (1) *is equal to the units digit of the nonnegative integer* $[10^k a]$ *for* $k = 1,$ $2, \cdots, m.$

Proof.   Every positive number of the form (1) is clearly a positive integer divided by 10^m (take a common denominator). That any such number a is less than 1 follows from the inequalities (cf. proof of the Theorem, § 1201, and Theorem II, § 314):

$$a \leq \frac{9}{10} + \cdots + \frac{9}{10^m} = \frac{9}{10}\left(1 + \cdots + \left(\frac{1}{10}\right)^{m-1}\right) = 1 - \left(\frac{1}{10}\right)^m < 1.$$

Conversely, let $p/10^q$ be a decimal fraction between 0 and 1, where p is a positive integer between 0 and 10^q. Then (§ 1201) p has a representation of the form $d_{-q+1}10^{q-1} + \cdots + d_0$. Division by 10^q, and a change in notation, gives a terminating decimal in the form (1). If

$$a = \frac{d_1}{10} + \frac{d_2}{10^2} + \cdots + \frac{d_m}{10^m} = \frac{e_1}{10} + \frac{e_2}{10^2} + \cdots + \frac{e_r}{10^r}.$$

where the d's and e's are decimal digits, $d_m \neq 0$, $e_n \neq 0$, and $m \leq r$, then multiplication by 10^r gives

$$d_1 10^{r-1} + d_2 10^{r-2} + \cdots + d_m 10^{r-m} = e_1 10^{r-1} + \cdots + e_r.$$

By the Theorem, § 1201, this representation is unique ($m = r$, and $d_k = e_k$ for $k = 1, 2, \cdots, m$). Finally, for any $k = 1, 2, \cdots, m$,

$$10^k a = 10^{k-1} d_1 + \cdots + 10 d_{k-1} + d_k + \frac{d_{k+1}}{10} + \cdots + \frac{d_m}{10^{m-k}}.$$

As in the proof of the Theorem, § 1201, d_k is the units digit of $[10^k a]$.

Example.   Find the terminating expansion with base six for the rational number whose decimal notation is 47/144.

Solution.   Since $144 = 2^4 \cdot 3^2$ we first multiply numerator and denominator of 47/144 by $3^2 = 9$ in order to get a denominator that is a power of 6: $\frac{47}{144} = \frac{423}{1296}$.

We next express the numerator 423 in terms of powers of 6:

$$\begin{array}{ll} 6)\overline{423} & \textit{Remainders} \\ 6)\ \overline{\ 70} & 3 \\ 6)\ \overline{\ 11} & 4 \\ 6)\ \overline{\ \ 1} & 5 \\ \overline{\ \ \ 0} & 1. \end{array}$$

In decimal notation, $423 = 1\cdot6^3 + 5\cdot6^2 + 4\cdot6 + 3$. In the system with base six, the original fraction takes the form

$$\frac{1,543}{10,000} = .1543.$$

1204. INCREASING SEQUENCES

The concepts of *finite sequence, infinite sequence, decreasing sequence*, and *strictly decreasing sequence* were introduced in §§ 313 and 316. In the present chapter we shall be interested in reversing inequalities, and in studying *increasing sequences*, as defined below. It should be noted that decreasing and increasing sequences are special instances of the more general notions of decreasing and increasing functions discussed in § 1003.

In the remaining sections of this book the single word *sequence*, unless modified by the adjective *finite*, should be interpreted to mean *infinite sequence*. All sequences under consideration in this chapter are *infinite sequences of real numbers*.

Definition I. *A sequence $\{a_n\}$ is **increasing**† if and only if $a_n \leq a_{n+1}$ for every natural number n. A sequence $\{a_n\}$ is **strictly increasing** if and only if $a_n < a_{n+1}$ for every n. A sequence is **bounded, bounded above**, or **bounded below**, if and only if the set of all of its values, or terms, is bounded, bounded above, or bounded below, respectively.*

Note. An increasing sequence is always bounded below by its first term. Therefore, an increasing sequence is bounded if and only if it is bounded above.

Example 1. Of the sequences of Examples 2, § 313, (6) and (8) are increasing, but only (6) is strictly increasing; (7), (8), (9), and (11) are bounded; all are bounded below, but (6) and (10) are not bounded above.

Examples 2. The following are examples of strictly increasing bounded sequences:

(1) $.3, .33, .333, \cdots, \frac{1}{3}\left\{1 - \left(\frac{1}{10}\right)^n\right\}, \cdots;$

(2) $.1, .14, .141, .1414, .14141, .141414, \cdots, a_n, \cdots,$

† An increasing sequence is also called a **monotonically increasing sequence**, or a **non-decreasing sequence**.

$$\text{where } a_n = \begin{cases} \dfrac{1}{10} + \dfrac{41}{990}\left\{1 - \left(\dfrac{1}{10}\right)^{n-1}\right\} \text{ if } n \text{ is odd, and} \\[2ex] \dfrac{14}{99}\left\{1 - \left(\dfrac{1}{10}\right)^{n}\right\} \text{ if } n \text{ is even.} \end{cases}$$

Definition II. *A bounded increasing sequence is said to* **converge,** *or be* **convergent,** *and to have as* **limit** *the supremum or least upper bound of its values. This is written:*

(3) $\lim\limits_{n \to +\infty} a_n = a$, *or $a_n \to a$ as $n \to +\infty$,*

where

(4) $a \equiv \sup \{a_1, a_2, \cdots, a_n, \cdots\}.$

An increasing sequence that is not bounded is said to **diverge,** *or be* **divergent;** *it is also said to fail to converge or be convergent.*

The justification for the terminology and notation of the preceding definition is given by the following theorem.

   Theorem. *If $\{a_n\}$ is a bounded increasing sequence, and if $a \equiv \sup \{a_1, a_2, \cdots, a_n, \cdots\}$, then*

(5) $a_n \leqq a$, *for* $n = 1, 2, \cdots$.

If, in addition, $\{a_n\}$ is strictly increasing, then

(6) $a_n < a$, *for* $n = 1, 2, \cdots$.

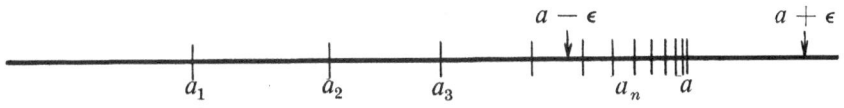

FIG. 1201

Furthermore, if $\{a_n\}$ is a bounded increasing sequence and if ϵ is an arbitrary positive number, there exists a corresponding natural number N such that the inequality $n > N$ implies

(7) $a_n > a - \epsilon$,

and therefore, in conjunction with (5):

(8) $a - \epsilon < a_n < a + \epsilon$, *or* $|a_n - a| < \epsilon$.

(Cf. Fig. 1201.)

   Proof. The inequality (5) follows from the fact that a is an upper bound of the set of all terms of the sequence. To prove (6), assume that equality holds for some term: $a_m = a$. Then for $n = m + 1$, $a_n = a_{m+1} > a_m = a$, in contradiction to (5). Since a is the *least* upper bound of $\{a_1, a_2, \cdots\}$, and $a - \epsilon < a$, $a - \epsilon$ is not an upper bound of $\{a_1, a_2, \cdots\}$. Consequently, there must exist a term a_N greater than $a - \epsilon$: $a_N > a - \epsilon$. Therefore, $n > N$ implies $a_n \geqq a_N$ and hence $a_n > a - \epsilon$. The equivalence of the inequalities in (8) is a consequence of part IV of Theorem I, § 804, with $x = a_n - a$.

Example 3. Show that the sequence

(9)
$$\frac{1}{2}, \frac{2}{3}, \frac{3}{4}, \cdots, \frac{n}{n+1}, \cdots$$

is strictly increasing and bounded, and find its limit.

Solution. If $a_n \equiv n/(n + 1)$, then $a_{n+1} = (n + 1)/(n + 2)$ and the inequality $a_n < a_{n+1}$ is equivalent to the inequality

$$\frac{n}{n+1}(n + 1)(n + 2) < \frac{n+1}{n+2}(n + 1)(n + 2),$$

or $n^2 + 2n < n^2 + 2n + 1$, which is true for all n. This establishes the fact that the sequence (9) is strictly increasing. It is obvious, since $n < n + 1$, that 1 is an upper bound for the terms of (9). To show that 1 is the *least* upper bound it is necessary only to show that no number *less* than 1 can be an upper bound. Accordingly, assume that b is a number less than 1 that is an upper bound of (9). Our objective is to find a positive integer n such that $n/(n + 1) > b$, and thus find a contradiction to the assumption that $b < 1$. The inequality $n/(n + 1) > b$ is equivalent to $n > nb + b$ which, in turn, is equivalent to $n(1 - b) > b$. Since $1 - b > 0$, the solutions of this last inequality are all positive integers n such that $n > b/(1 - b)$. Since such positive integers exist (Corollary I to Theorem I, § 903) the desired contradiction is obtained.

Example 4. Prove that if $\{m_n\}$ is any strictly increasing sequence of positive integers, then $m_n \geqq n$ for every n. Conclude that every strictly increasing sequence of integers is unbounded.

Solution. Let $P(n)$ be the proposition $m_n \geqq n$. Then $P(1)$ is true since 1 is the least positive integer. Assuming $P(n)$ to be true for the particular positive integer n, we wish to establish the truth of $P(n + 1)$; that is, we wish to show that the inequality $m_n \geqq n$ implies the inequality $m_{n+1} \geqq n + 1$. Since $\{m_n\}$ is strictly increasing, $m_{n+1} \geqq m_n + 1 \geqq n + 1$. Finally, if $\{a_n\}$ is a strictly increasing sequence of integers, let $b_n \equiv a_n - a_1 + 1$. Then $\{b_n\}$ is a strictly increasing sequence of *positive* integers, and $b_n \geqq n$. Therefore, $a_n \geqq n + a_1 - 1$, and since the set of all positive integers is unbounded above, so is the sequence $\{a_n\}$.

1205. INFINITE SERIES OF NONNEGATIVE TERMS

Let $\{a_n\}$ be a sequence of nonnegative real numbers, and form a new sequence called the **sequence of partial sums** of $\{a_n\}$:

(1)
$$s_n = a_1 + a_2 + \cdots + a_n, n = 1, 2, \cdots$$

The original sequence. then, is called an **infinite series**, or **series**, and is denoted Σa_n, or:

$$(2) \qquad \sum_{n=1}^{+\infty} a_n = a_1 + a_2 + \cdots + a_n + \cdots.$$

The **terms** of a series Σa_n are the numbers a_n, $n = 1, 2, \cdots$. Since $a_n \geq 0$ for every n, $s_{n+1} \geq s_n$ for every n, and the sequence $\{s_n\}$ of partial sums of Σa_n is an increasing sequence. If $\{s_n\}$ is bounded, its limit $s \equiv \sup \{s_1, s_2, \cdots\}$ is called the **sum** of the infinite series Σa_n, with the notation:

$$(3) \qquad s = \sum a_n = \sum_{n=1}^{+\infty} a_n = a_1 + a_2 + \cdots + a_n + \cdots.$$

In this case the series Σa_n is said to be **convergent**, and to **converge** to s. If $\{s_n\}$ is unbounded the series Σa_n is said to **diverge** or to be **divergent**.

For the sake of emphasis, let us extract the following statement from the preceding paragraph:

Theorem I. *A series of nonnegative terms converges if and only if its partial sums are bounded.*

A useful basic theorem concerning series of nonnegative terms is the comparison test.

Theorem II. Comparison Test. *If Σa_n and Σb_n are series of nonnegative terms, if $a_n \leq b_n$ for every n, and if Σb_n converges, then Σa_n converges too. In this case, if $s \equiv \Sigma a_n$ and $t \equiv \Sigma b_n$, then $s \leq t$.*

Proof. Let $\{s_n\}$ and $\{t_n\}$ be the sequences of partial sums of the series Σa_n and Σb_n, respectively. The equality $a_n \leq b_n$ implies immediately the inequality $s_n \leq t_n$, for all n. Therefore, if Σb_n converges, $\{t_n\}$ is bounded and $\{s_n\}$ must be bounded too. Consequently, Σa_n converges. To show that $s \leq t$, assume $s > t$, and let $\epsilon = s - t > 0$. By the Theorem, § 1204, applied to the sequence $\{s_n\}$ there exists a number N such that $n > N$ implies $s_n > s - \epsilon = t$ and hence $t_n \geq s_n > t$. But this is in contradiction to the inequality $t_n \leq t$ (cf. (5), § 1204).

Example. Show that the series

$$(4) \qquad \frac{1}{1^2} + \frac{1}{2^2} + \frac{1}{3^2} + \cdots + \frac{1}{n^2} + \cdots$$

converges.

Solution. Consider the two related series:

$$(5) \qquad \sum a_n = \frac{1}{2^2} + \frac{1}{3^2} + \cdots + \frac{1}{(n+1)^2} + \cdots,$$

$$(6) \qquad \sum b_n = \frac{1}{1 \cdot 2} + \frac{1}{2 \cdot 3} + \cdots + \frac{1}{n(n+1)} + \cdots.$$

These are series of positive terms, and $a_n < b_n$ for every n. Furthermore,

$$b_1 + \cdots + b_n = \left(\frac{1}{1} - \frac{1}{2}\right) + \left(\frac{1}{2} - \frac{1}{3}\right) + \left(\frac{1}{3} - \frac{1}{4}\right) + \cdots + \left(\frac{1}{n} - \frac{1}{n+1}\right)$$

$$= \frac{1}{1} - \frac{1}{n+1} = \frac{n}{n+1}.$$

Since the sequence of partial sums of $\sum b_n$ is that of Example 3, § 1204, the series $\sum b_n$ converges. Consequently, by the comparison test (Theorem II), the series $\sum a_n$ converges. If u denotes any upper bound for the partial sums of series (5), $u + 1$ is an upper bound for the partial sums of the series (4). Since these partial sums are bounded, the series (4) must converge.

1206. GEOMETRIC SERIES

A series of the form

(1) $$a + ar + ar^2 + \cdots + ar^{n-1} + \cdots$$

is called a **geometric series** with first term a and common ratio r. Although much interest attaches to the series (1) for all a and for all r between -1 and 1, we shall be interested in this series only for the following values of a and r: $a > 0, 0 < r < 1$.

Theorem I. *If $a > 0$ and $0 < r < 1$, the geometric series* (1) *converges to the sum $a/(1 - r)$:*

(2) $$\frac{a}{1 - r} = a + ar + ar^2 + \cdots + ar^{n-1} + \cdots.$$

Proof. If s_n designates the partial sum of the first n terms of (1) then by the formula for the sum of a *finite* geometric sequence (Theorem II, § 314):

(3) $$s_n = a(1 + r + \cdots + r^{n-1}) = a\frac{1 - r^n}{1 - r} < \frac{a}{1 - r},$$

and we see that $a/(1 - r)$ is an upper bound for $\{s_1, s_2, \cdots\}$. To show that $a/(1 - r)$ is the *least* upper bound of this set, assume that there exists an upper bound b *less* than $a/(1 - r)$ and define $\epsilon \equiv \dfrac{a}{1 - r} - b$ so that $b = \dfrac{a}{1 - r} - \epsilon$.

Then for every positive integer n:

(4) $$a\frac{1 - r^n}{1 - r} \leq \frac{a}{1 - r} - \epsilon, \quad \text{or} \quad r^n \geq \frac{\epsilon(1 - r)}{a}$$

This means that the set $\{r, r^2, r^3, \cdots, r^n, \cdots\}$ has a *positive lower bound*. Let ρ designate the *greatest* lower bound of this set, so that $r^n \geq \rho > 0$ for every positive integer n. This implies that

(5) $$r^{n-1} \geq \frac{\rho}{r} > \rho > 0,$$

for every positive integer n, and as a consequence, $r^n \geq \rho/r$. But this means that ρ/r is a lower bound of the set $\{r, r^2, r^3, \cdots, r^n, \cdots\}$ *greater* than the *greatest* lower bound! With this contradiction the proof is complete.

Example 1. Find the sum of the geometric series

$$98 + 42 + 18 + \tfrac{54}{7} + \cdots + 98(\tfrac{3}{7})^{n-1} + \cdots.$$

Solution. Since $a = 98$ and $r = \tfrac{3}{7}$, the sum is

$$\frac{98}{1 - \tfrac{3}{7}} = \frac{7}{4} \cdot 98 = \frac{343}{2} = 171\tfrac{1}{2}.$$

A useful extension of the result in Theorem I is the following:

Theorem II. *If c is any number, if $a > 0$ and if $0 < r < 1$, then the series*

(6) $$c + a + ar + ar^2 + \cdots + ar^{n-2} + \cdots \quad (n > 1)$$

converges to the sum $c + \dfrac{a}{1 - r}$.

Proof. If t_n denotes the partial sum of $n + 1$ terms of (6), and if s_n denotes the partial sum of n terms of (1), as in the proof of Theorem I, then $t_n = c + s_n$ for $n = 1, 2, \cdots$. If $s = a/(1 - r)$, then $s_n < s$ for every n, $t_n < c + s$ for every n, and therefore the series (6) converges and has a sum $t \leq c + s$. On the other hand, the inequality $t_n < t$ implies $s_n < t - c$ for every n, and hence $s \leq t - c$, or $t \geq c + s$. Therefore $t = c + s$, as desired.

Example 2. The series

$$\tfrac{57}{2} + 98 + 42 + 18 + \tfrac{54}{7} + \cdots,$$

with terms continuing as in Example 1 converges and has sum equal to $\tfrac{57}{2} + \tfrac{343}{2} = 200$.

1207. CANONICAL DECIMAL EXPANSIONS

For reasons outlined in § 1203 we shall limit our considerations of decimal expansions to those for positive numbers less than 1. We start by defining the symbol $.d_1 d_2 \cdots d_n \cdots$:

Definition I. *The symbol* $.d_1 d_2 \cdots d_n \cdots$, *where the **coefficient** or **digit** d_n is a nonnegative integer less than 10, for $n = 1, 2, \cdots$, is called a **decimal expansion**, and is an abbreviation for the infinite series:*

(1) $$.d_1 d_2 \cdots d_n \cdots \equiv \frac{d_1}{10} + \frac{d_2}{10^2} + \cdots + \frac{d_n}{10^n} + \cdots.$$

If only a finite number of the digits are nonzero the decimal expansion (1) *is **terminating** (cf. § 1203); otherwise, if infinitely many digits are nonzero, the decimal expansion* (1) *is **nonterminating**.*

Our first main problem is to show how a decimal expansion can be obtained for an arbitrary given positive number less than 1, and to prove that this decimal expansion converges to the given number.

Theorem I. *If a is any positive number less than 1: $0 < a < 1$, and if d_n is the units digit of the nonnegative integer $[10^n a]$, for $n = 1, 2, \cdots$, then the infinite series $.d_1 d_2 \cdots d_n \cdots$ converges to a:*

(2)
$$a = .d_1 d_2 \cdots d_n \cdots .$$

If s_n denotes the partial expansion or partial sum of (2): $s_n \equiv .d_1 d_2 \cdots d_n$, then s_n satisfies the inequalities

(3)
$$a - \frac{1}{10^n} < s_n \leq a,$$

for $n = 1, 2, \cdots$.

Proof. Let $P(n)$ designate the assertion of the inequalities (3) for the natural number n. Then $P(1)$ is true since $0 < a < 1$ implies $0 < 10a < 10$, and hence $0 \leq [10a] \leq 10a < 10$. Since, for any real number x, $x - 1 < [x] \leq x$, $10a - 1 < d_1 = [10a] \leq 10a$, and hence $a - \frac{1}{10} < s_1 = \frac{d_1}{10} \leq a$. Assuming the truth of $P(n)$ for a particular natural number n, we wish to establish the truth of $P(n + 1)$. Since $10^{n+1} s_n$ is an integer divisible by 10, the units digit of $[10^{n+1} a]$ is the same as the units digit of $[10^{n+1}(a - s_n)]$. Furthermore, from the inequalities $a - \frac{1}{10^n} < s_n \leq a$, we immediately deduce $0 \leq 10^{n+1}(a - s_n) < 10$, so that $d_{n+1} = [10^{n+1}(a - s_n)]$. Again using $x - 1 < [x] \leq x$, we conclude $10^{n+1}(a - s_n) - 1 < d_{n+1} \leq 10^{n+1}(a - s_n)$, or $a - \frac{1}{10^{n+1}} < s_n + \frac{d_{n+1}}{10^{n+1}} \leq a$. With this, the truth of $P(n + 1)$ is established. Therefore (3) holds for all natural numbers n. Since inequalities (3) imply in particular that the partial sums of (2) are bounded, the convergence of $.d_1 d_2 \cdots d_n \cdots$ follows. We also know that if s is the sum of the series $.d_1 d_2 \cdots d_n \cdots$, then (from (3) again) $s \leq a$. It only remains to show that the inequality $s < a$ is impossible and that the equality of (2) must obtain. Assuming $s < a$ would force us, as a consequence of the first inequality of (3) and the inequality $s_n \leq s$ of (5), § 1204, to the conclusion that $a - \frac{1}{10^n} < s$ for all natural numbers n. But this would mean that $\frac{1}{10^n}$ is always greater than the positive number $a - s$, and consequently that every natural number n must satisfy the inequalities $n < 10^n < 1/(a - s)$. Since the natural numbers are not bounded (cf. § 903), the desired contradiction has been reached and the proof is complete.

We have found in §§ 1201 and 1203 that any *terminating* decimal expansion is unique. To what extent does this uniqueness carry over to decimal expansions in general? The answer (such as it is) is "most of the way." The following example indicates the nature of the nonuniqueness when it occurs.

Example 1. Show that

(4) $$.24000 \cdots = .23999 \cdots,$$

where the dots indicate repeated 0's and 9's.

Solution. By Theorem II, § 1206, since the second decimal expansion of (4) is an infinite series of the form $c + a + ar + \cdots + ar^{n-2}$, where $c = .23$, $a = .009$, and $r = .1$, it has a sum equal to $c + a/(1 - r) = .23 + .009/.9 = .24$.

The same technique as that used in the solution of Example 1 shows that any decimal expansion with solidly repeating 9's is equal to a terminating decimal:

$$.d_1 d_2 \cdots d_n 999 \cdots = .d_1 d_2 \cdots (d_n + 1),$$

where $d_n < 9$. If we rule out such atypical expansions, we can establish uniqueness. We begin with the definition:

Definition II. *The decimal expansion*

(5) $$.d_1 d_2 \cdots d_n \cdots$$

is **canonical** *if and only if the inequality $d_n < 9$ holds for infinitely many subscripts n.*

Concerning the expansions given by Theorem I we have the theorem:

Theorem II. *If $0 < a < 1$ and if d_n is the units digit of $[10^n a]$, for $n = 1$, $2, \cdots$, then the expansion $a = .d_1 d_2 \cdots d_n \cdots$ is canonical.*

Proof. If $a = .d_1 d_2 \cdots d_n 999 \cdots$, then $a = .d_1 d_2 \cdots (d_n + 1)$, and a is equal to a decimal rational. In this case, by § 1203, its decimal expansion, as prescribed by the rule for d_n stated above, terminates. This is the desired contradiction.

A converse to Theorem II is the following:

Theorem III. *If $.d_1 d_2 \cdots d_n \cdots$ is a canonical decimal expansion, if not every digit d_n is equal to 0, and if a is its sum, then $0 < a < 1$.*

Proof. The inequality $a > 0$ is obvious. Let m be the least subscript such that $d_m \leq 8$. Then $a \leq .d_1 d_2 \cdots d_m 999 \cdots$. This last decimal expansion, by Theorem II, § 1206, with $c = .d_1 d_2 \cdots d_m$, $a = .00 \cdots 09$, and $r = .1$, is equal to $.d_1 d_2 \cdots (d_m + 1) \leq .d_1 d_2 \cdots d_{m-1} 9 < 1$.

Every positive number less than 1 has a unique canonical decimal expansion. In precise terms:

Theorem IV. *If $0 < a < 1$, and if $.d_1 d_2 \cdots d_n \cdots$ and $.e_1 e_2 \cdots e_n \cdots$ are* *two canonical decimal expansions whose sums are both equal to a:*

(6) $a = .d_1 d_2 \cdots d_n \cdots = .e_1 e_2 \cdots e_n \cdots ,$

then $d_n = e_n$ for every $n = 1, 2, \cdots$.

Proof. Let k be the least subscript such that $d_k \neq e_k$, and assume that the notation is such that $d_k < e_n$. Now let m be the least subscript greater than k such that $d_m \leq 8$. Then, by the same process as that used in the proof of Theorem III, we see that $.d_1 d_2 \cdots d_k \cdots d_{m-1} d_m \cdots \leq .d_1 d_2 \cdots$ $d_k \cdots d_{m-1}9 < .d_1 d_2 \cdots d_{k-1}(d_k + 1) \leq .e_1 e_2 \cdots e_{k-1}e_k \leq .e_1 e_2 \cdots e_n \cdots .$ In other words, $a < a$. This contradiction establishes the desired uniqueness.

As a corollary to Theorem IV we have:

Theorem V. *There is a one-to-one correspondence between the points of the half-open interval $[0, 1)$ (that is, the numbers a such that $0 \leq a < 1$) and all canonical decimal expansions $.d_1 d_2 \cdots d_n \cdots$ (that is, all sequences of digits $d_n = 0, 1, 2, \cdots , 9$ such that infinitely many of the digits are different from 9).*

Proof. Every canonical expansion $.d_1 d_2 \cdots d_n \cdots$ converges to a number a such that $0 \leq a < 1$, and every number a such that $0 \leq a < 1$ determines a unique canonical expansion $.d_1 d_2 \cdots d_n \cdots$ that converges to a.

1208. INFINITE DIVISION ALGORITHM

In order to obtain the canonical decimal expansion for a rational number, like 9/13 (decimal notation), the customary procedure is to divide 9 by 13 in much the same style as one would divide a positive integer by a divisor, except that in the case of 9/13, the process continues without end. Does this actually give the canonical expansion, and if so why? Let us take a good look at this example and, in this way, see why the method is perfectly valid in general. The division process, or algorithm, for 9/13 is:

```
              69230769 · · ·
     13 )9.00000000
          78
          120
          117
            30
            26
            40
            39
            100
             91
             90
             78
            120.
```

Let us look at the first five digits, d_1, d_2, d_3, d_4, and d_5 of the canonical expansion, and see that they are, indeed, 6, 9, 2, 3, and 0, respectively. To find d_1, we first seek $[10 \cdot \frac{9}{13}] = [\frac{90}{13}] = [6\frac{12}{13}] = 6$. Hence d_1 is equal to the units digit of $6 = 6$. To find d_2 we compute $[10^2 \cdot \frac{9}{13}] = [\frac{900}{13}] = [69\frac{3}{13}] = 69$, whose units digit is 9. For d_3, $[10^3 \cdot \frac{9}{13}] = [\frac{9000}{13}] = [692\frac{4}{13}] = 692$, with a units digit of 2. Finally, for d_4 and d_5, we have the units digits of $[\frac{90000}{13}] = [6923\frac{1}{13}]$ and $[\frac{900000}{13}] = [69230\frac{10}{13}] = 69230$, so that $d_4 = 3$ and $d_5 = 0$.

The reader may wish to carry through the details on other examples in order to convince himself that the infinite division algorithm illustrated by the example of 9/13 is universally justified.

1209. REPEATING DECIMALS

In the example of § 1208, it is clear after momentary inspection that the digits $692 \cdots$ have begun to repeat themselves in a cycle of 6, and that the decimal expansion of 9/13 is

(1) $\frac{9}{13} = 692307692307692307692307 \cdots$

It does not take long to discover why such a cyclic repetition *must* take place if we look at the successive remainders after division: 12, 3, 4, 1, 10, 9, 12, 3, 4, $\cdots$. Unless the remainder turns out to be 0—in which case the decimal terminates—there are finitely many nonzero remainders possible, namely 1, 2, 3, $\cdots$, 12. Therefore the length of any possible sequence of divisions without a repeated remainder can be at most 12. Thus a repeated remainder must occur, and the division cycle repeats itself.

Let us try another example, 87/110:

$$
\begin{array}{r}
.7909 \cdots \\
\overline{87.0} \\
77\;0 \\
\overline{10\;00} \\
9\;90 \\
\overline{1000} \\
990. \\
\end{array}
$$

It is evident from this division that

(2) $\frac{87}{110} = .79090909090 \cdots$

We notice that the cyclic repetition in this case starts with the second digit (not the first) and continues in a cycle of 2.

The restriction of our considerations to positive numbers less than 1 has been for convenience of notation only, as brief reflection, and an illustrative example or two, will show. Division by a power of 10 in the denominator serves to transform a fraction greater than 1 to one less than 1; a corresponding shift of the decimal point completes the story. To illustrate with 87/11:

$$
\begin{array}{r}
7.909\cdots \\
11\overline{\smash{\big)}\,87} \\
77 \\
\hline
100 \\
99 \\
\hline
100 \\
99. \\
\hline
\end{array}
$$

Comparison with the preceding example shows that no further justification is needed.

If we define a **repeating decimal** to be either a terminating decimal (with repeating 0's) or a nonterminating decimal with a cyclic repetition of a sequence of digits, and if we include negative numbers by simply attaching negative signs to the decimal expansions of their absolute values, we have the result:

Theorem I. *Every rational number has a repeating decimal expansion. Equivalently, every nonrepeating decimal expansion converges to an irrational number.*

The converse of this is also true:

Theorem II. *Every repeating decimal expansion converges to a rational number. Equivalently, every irrational number has a nonrepeating decimal expansion.*

Proof. We may clearly limit our considerations to canonical decimal expansions for positive numbers less than 1. Terminating expansions were considered in § 1203. Nonterminating repeating decimals have the form:

$$(3) \qquad .d_1 d_2 \cdots d_k e_1 e_2 \cdots e_p e_1 e_2 \cdots e_p e_1 e_2 \cdots e_p \cdots .$$

Since any partial sum s_n of (3), for $n > k$, satisfies an inequality of the form

$$s_{k+mp} \leq s_n \leq s_{k+(m+1)p}, \quad \text{where} \quad k + mp \leq n \leq k + (m + 1)p,$$

it is sufficient to consider the limit of s_{k+mp}, as $m \to +\infty$. (This is true since $\sup \{s_{k+mp}\} \leq \sup \{s_n\} \leq \sup \{s_{k+(m+1)p}\} = \sup \{s_{k+mp}\}$.) Any sum s_{k+mp} has the form

$$(4) \qquad s_{k+mp} = c + a + ar + \cdots + ar^{m-1},$$

where $c = .d_1 d_2 \cdots d_k$, $a = 10^{-k}(.e_1 e_2 \cdots e_p)$, and $r = 10^{-p}$. Therefore, by Theorem II, § 1206,

$$s = \lim_{m \to +\infty} s_{k+mp} = c + \frac{a}{1 - r},$$

which is a rational number. This completes the proof.

We illustrate Theorem II by taking the decimal expansions of 9/13 and 87/110, obtained above, and "rediscovering" the original fractions that produced them.   In the case of 9/13, with the notation of the proof of Theorem II, $k = 0, p = 6, a = .692307$, and $r = 10^{-6}$, so that the sum of the infinite series in (1) is $a + ar + ar^2 + \cdots + ar^{m-1} + \cdots = a(1-r) = .692307/.999999 = 692307/999999 = 9/13$.   In the case of 87/110, $k = 1$, $p = 2, c = .7, a = .090$, and $r = 10^{-2}$, and the infinite series in (2) becomes

$$.7 + .090 + (.090)10^{-2} + (.090)10^{-4} + \cdots$$

$$= .7 + \frac{.090}{1 - 10^{-2}} = \frac{7}{10} + \frac{9}{99} = \frac{7}{10} + \frac{1}{11} = \frac{87}{110}.$$

Notation.   In the examples and exercises that follow, we shall adopt the convention of placing a horizontal line or bar over the digits of a repeating cycle.   Thus, in the examples already considered, we could write:

$$\tfrac{9}{13} = .\overline{692307}, \quad \tfrac{87}{110} = .7\overline{90}, \quad \text{and} \quad \tfrac{87}{11} = 7.\overline{90}.$$

Example 1.   Show that $\sqrt{2} \neq 1.\overline{4}$.

Solution.   Let $b = 1.\overline{4} = 1.4141414\cdots$.   Then, with $a = 1.4$ and $r = 10^{-2}$, $b = \dfrac{a}{1-r} = \dfrac{1.4}{.99} = \dfrac{140}{99}$, and b is a rational number.   On the other hand (§ 905), $\sqrt{2}$ is irrational, so $\sqrt{2} \neq b$.

For further illustrative material we consider two examples where the base is not ten:

Example 2.   Find the quinary expansion of the fraction 2/3 and check.

Solution.   Repetitive division gives

$$
\begin{array}{r}
.313\cdots \\
3\overline{)2.0000} \\
14 \\
\hline
10 \\
3 \\
\hline
20 \\
14 \\
\hline
\cdots
\end{array}
$$

The process has started to repeat, so that the quinary expansion desired is .3131313131 $\cdots$ or $.\overline{31}$.   To check, we use the formula $a/(1-r)$, where $a = .31$ and $r = .01$, with the result: $\dfrac{.31}{.44} = \dfrac{31}{44} = \dfrac{2\cdot13}{3\cdot13} = \dfrac{2}{3}$.

Example 3.   What rational number has the binary expansion $.1\overline{100}$?

First solution. Using decimal notation we rewrite the binary expansion in the form

$$\tfrac{1}{2} + \tfrac{1}{4} + \tfrac{1}{4}(\tfrac{1}{8}) + \tfrac{1}{4}(\tfrac{1}{8})^2 + \cdots,$$

which has the sum $c + \dfrac{a}{1-r}$, where $c = \tfrac{1}{2}$, $a = \tfrac{1}{4}$, and $r = \tfrac{1}{8}$, or:

$$\frac{1}{2} + \frac{\tfrac{1}{4}}{1 - \tfrac{1}{8}} = \frac{1}{2} + \frac{1}{4} \cdot \frac{8}{7} = \frac{1}{2} + \frac{2}{7} = \frac{11}{14}.$$

Second solution. With binary notation, the above evaluation takes the form:

$$.1 + \frac{.0100}{1 - .001} = \frac{1}{10} + \frac{10}{111} = \frac{111 + 100}{1110} = \frac{1011}{1110}.$$

This result is in agreement with the first solution since, in the decimal system the final numerator is equal to $8 + 2 + 1 = 11$ and the denominator is equal to $8 + 4 + 2 = 14$.

1210. THE SQUARE ROOT ALGORITHM

In this section we shall examine the standard procedure, or algorithm, for obtaining the decimal expansion for the square root of a given positive number. Our main objective will be to establish the validity of this square root algorithm. For simplicity of notation we shall limit our theoretical discussions to positive numbers less than 1, but for illustrative material we shall feel free to use arbitrary positive numbers, whether greater or less than 1. For any number greater than 1 we have only to divide by a power of 100 and multiply the answer by the same power of 10. For example, if the expansion $\sqrt{.02} = .141421 \cdots$ is justified, then we can infer $\sqrt{2} = 1.41421 \cdots$, $\sqrt{200} = 14.1421 \cdots$, etc.

Let us recall the square root algorithm through an example, $\sqrt{.02}$. If we wish to determine the first 6 digits (to the right of the decimal point), we mark off 12 digits in pairs:

$$.02\ 00\ 00\ 00\ 00\ 00.$$

The next step is to find the largest integer whose square does not exceed the number represented by the first pair, in this case 2, and proceed thus:

$$
\begin{array}{r}
.1 \\
\hline
\sqrt{.02\ 00\ 00\ 00\ 00\ 00.} \\
1\ 00 \\
\hline
1\ 00
\end{array}
$$

Now double the 1 found on top, and determine the largest digit d such that

$(10 \cdot 2 \cdot 1 + d) d \leq 100$. This digit is 4, and it is placed on the top row:

```
              .1  4
       √‾.02 00 00 00 00 00,
            1  00
24          1  00
 4             96
96           4  00
```

Continuing in this manner, we have

```
24                  .1  4  1  4  2  1
 4           √‾.02 00 00 00 00 00 00
96                1  00
       281        1  00
         1           96
       281   2824    4  00
               4     2  81
             11296   1  19 00
28282                1  12 96
    2                   6  04 00
56564                   5  65 64
         282841           38  36 00
              1           28  28 41
         282841
```

We conclude: to 6 decimal places $\sqrt{.02} = .141421$. Is this correct? If so, why?

Let $.d_1 d_2 d_3 d_4 \cdots d_{2n-1} d_{2n} \cdots$ be the canonical decimal expansion of a positive number a less than 1. If r_1 is the largest digit whose square does not exceed $10d_1 + d_2$, then

$$r_1^2 \leq 10d_1 + d_2 < (r_1 + 1)^2.$$

This is equivalent to $(.r_1)^2 \leq .d_1 d_2 < (.(r_1 + 1))^2$. Consequently, $(.r_1)^2 \leq a < (.(r_1 + 1))^2$. (If *equality* held: $a = (.(r_1 + 1))^2$, then a would be a terminating decimal and hence *equal* to $.d_1 d_2$—which is *less* than $(.(r_1 + 1))^2$.) In other words,

$$.r_1 \leq \sqrt{a} < .(r_1 + 1),$$

and r_1 is the correct first digit of $\sqrt{a}$.

We next proceed by doubling r_1† and determining the largest digit r_2 such

---

† The reader should observe that the doubling of r_1 is directly related to the coefficient 2 in the middle term of the binomial expansion with exponent 2: $(a + b)^2 = a^2 + 2ab + b^2$.

that $(10 \cdot 2r_1 + r_2)r_2$ does not exceed the difference between $10^3d_1 + 10^2d_2 + 10d_3 + d_4$ and $100r_1^2$:

$$(20r_1 + r_2)r_2 \leq 10^3d_1 + 10^2d_2 + 10d_3 + d_4 - 100r_1^2$$

$$< (20r_1 + r_2 + 1)(r_2 + 1).$$

These inequalities are equivalent to:

$$\{10r_1 + r_2\}^2 \leq 10^3d_1 + 10^2d_2 + 10d_3 + d_4$$

$$< \{10r_1 + (r_2 + 1)\}^2,$$

or:

$$(.r_1r_2)^2 \leq .d_1d_2d_3d_4 < (.r_1(r_2 + 1))^2.$$

We therefore have, as before,

$$(.r_1r_2)^2 \leq a < (.r_1(r_2 + 1))^2,$$

or:

$$.r_1r_2 \leq \sqrt{a} < .r_1(r_2 + 1),$$

and r_2 is the correct second digit in the decimal expansion of $\sqrt{a}$.

We could now continue by mathematical induction, but since the notation is cumbersome and the ideas relatively elementary, we feel that it is more instructive to present the details for the third digit r_3, and leave the induction proof to any reader who is sufficiently interested.

Accordingly, we double the number $10r_1 + r_2$, multiply by 10, and seek the largest digit r_3 such that the product $(200r_1 + 20r_2 + r_3)r_3$ does not exceed $100\{10^3d_1 + 10^2d_2 + 10d_3 + d_4 - (10r_1 + r_2)^2\} + 10d_5 + d_6$—that is, the preceding difference shifted by means of a factor of 100, with two more digits included. The requirement imposed on r_3 takes the form of the inequalities:

$$(100r_1 + 10r_2 + r_3)^2 \leq 10^5d_1 + 10^4d_2 + \cdots + 10d_5 + d_6$$

$$< (100r_1 + 10r_2 + (r_3 + 1))^2,$$

or: $(.r_1r_2r_3)^2 \leq .d_1d_2 \cdots d_6 < (.r_1r_2(r_3 + 1))^2.$ From this we conclude:

$$.r_1r_2r_3 \leq \sqrt{a} < .r_1r_2(r_3 + 1),$$

and r_3 is the correct third digit in the decimal expansion of $\sqrt{a}$.

Example. Find the binary expansion of $\sqrt{2}$ to 8 digits (7 to the right of the "binary point"), and interpret the resulting inequalities in the decimal system.

Solution. In the binary system the number 2 takes the form 10, and the square root algorithm proceeds as follows (since, in the binary system, doubling merely involves attaching a 0, and the only digits are 0 and 1, the process is particularly simple):

$$
\begin{array}{r}
1.\ 0\ 1\ 1\ 0\ 1\ 0\ 1 \\
\sqrt{10.\ 00\ 00\ 00\ 00\ 00\ 00\ 00} \\
1 \\
\hline
1\ \ 00\ 00 \\
10\ 01 \\
\hline
1\ 11\ 00 \\
1\ 01\ 01 \\
\hline
1\ 11\ 00\ 00 \\
1\ 01\ 10\ 01 \\
\hline
1\ 01\ 11\ 00\ 00 \\
1\ 01\ 10\ 10\ 01
\end{array}
$$

In the decimal system the result just obtained can be expressed by the inequalities:

$$1 + \tfrac{1}{4} + \tfrac{1}{8} + \tfrac{1}{32} + \tfrac{1}{128} < \sqrt{2} < 1 + \tfrac{1}{4} + \tfrac{1}{8} + \tfrac{1}{32} + \tfrac{2}{128},$$

or

$$\tfrac{181}{128} < \sqrt{2} < \tfrac{182}{128} = \tfrac{91}{64}.$$

1211. EXERCISES

In Exercises 1–4, the number given is written in the decimal system. Rewrite it in the binary system; quaternary system; duodecimal system.

1. 100.   **2.** 144.

3. 255.   **4.** 999.

In Exercises 5–12, perform the indicated operation both in the given decimal system and in the system with indicated base, and check by converting your answer to decimal form.

5. $27 + 51$; base two.   **6.** $967 + 251$; base twelve.

7. $137 - 69$; base five.   **8.** $311 - 149$; base seven.

9. 87×38; base two.   **10.** 113×217; base three.

11. $\frac{312}{24}$; base four.   **12.** $\frac{943}{41}$ base nine.

⋆13. Prove that any positive integer is congruent modulo nine to the sum of its decimal digits. This principle provides a simple means of checking arithmetic operations, known as *casting out nines*. To illustrate: to check $87 \times 35 = 3045$, we write (modulo 9): $87 \equiv 8 + 7 = 15 \equiv 1 + 5 = 6$ (modulo 9), $35 \equiv 3 + 5 = 8$ (modulo 9), so that $87 \times 35 \equiv 6 \times 8 = 48 \equiv 4 + 8 = 12 \equiv 1 + 2 = 3$ (modulo 9). On the other hand, $3045 \equiv 3 + 0 + 4 + 5 \equiv 3$ (modulo 9). Show that the method of casting out nines is not an infallible check. Discuss casting out threes and casting out elevens, with examples, in the decimal system.

14. Prove that any number represented by a terminating expansion in the system with base m is also represented by a terminating expansion in any system with base mn, where m and n are arbitrary integers greater than 1.

In Exercise 15–18, the fraction is given in decimal notation. Write a terminating expansion for the given number in the specified system.

15. $\frac{2877}{3125}$; quinary.   **16.** $\frac{401}{512}$; binary.

17. $\frac{401}{512}$; quaternary.   **18.** $\frac{13}{18}$; duodecimal.

In Exercises 19–22, find the rational number equal to the given repeating decimal, where the decimal system is used throughout.   Check by applying division to your answer.

19. $.35272\overline{72}7.$ **20.** $.1296296\overline{296}.$

21. $.594\overline{05}940.$ **22.** $.692307\overline{692307}.$

In Exercises 23–26, write out the (repeating or terminating) expansion for the number one-third in the system with specified base.

23. Two.   **24.** Five.

25. Six.   **26.** Eleven.

In Exercise 27–30, write out the (repeating or terminating) expansion for the number three-sevenths in the system with specified base.

27. Two.   **28.** Three.

29. Eight.   **30.** Twelve.

31. Write in terms of decimal numerals the rational number represented by the repeating expansion $.011011\overline{011}$, if the base is two; three; six; ten; twelve.

32. The **period** of a repeating decimal is the number of digits in the repeating part of the decimal.   For example, the period of $1/3 = .3333 \cdots$ is 1, and the period of $1/7 = .142857$ is 6.   Prove that the repeating decimal for p/q has period less than q. Find the period π for every repeating (and nonterminating) decimal for $1/q$, for $1 < q < 40$.

In Exercises 33–36, find the expansion of $\sqrt{2}$ in the indicated system to the specified number of digits to the right of the base point.

33. Base three; five digits.   **34.** Base five; four digits.

35. Base six; three digits.   **36.** Base twelve; three digits.

1212. COUNTABLE SETS

In § 312, finite sets were defined as either empty or such that their numbers are in one-to-one correspondence with the natural numbers from 1 to n, for some n.   It was demonstrated in Theorem IV, § 312, that the points of a finite set cannot be put into one-to-one correspondence with those of a proper subset.   Since such a correspondence between the natural numbers and a proper subset is possible (Theorem V, § 312), the set $\mathcal{N}$ of all natural numbers is infinite.   This points up an important *distinction* between the finite and the infinite.   In contrast to this distinction, as we shall see presently, there is a close *similarity* between finite sets and certain infinite sets.   This similarity turns out to hold, for example, for the set $\mathcal{N}$ of all natural numbers, for the set $\mathcal{I}$ of all integers, and for the set $\mathcal{Q}$ of all rational numbers, but *not* for the set $\mathcal{R}$ of all real numbers.   The underlying idea is *counting*.   We formulate a definition and give some examples.   For a historical discussion, see E. T. Bell, *The Development of Mathematics* (New York, McGraw-Hill Book Co., Inc., 1940), p. 138.

Definition. *A set is **denumerable** or **countably infinite** if and only if its members can be put into a one-to-one correspondence with the natural numbers. A set is **countable** if and only if it is either finite or denumerable. Equivalently, a set is countable if and only if its members can be arranged in a sequence of distinct terms, finite or infinite: $a_1, a_2, a_3, \cdots, a_n, \cdots$, where $a_m \neq a_n$ whenever $m \neq n$. A set is **uncountable** if and only if it fails to be countable; that is, if and only if it is infinite and there exists no one-to-one correspondence between its members and the natural numbers.*

Example 1. The set of all even natural numbers is countable since the correspondence $n \leftrightarrow 2n$ is one-to-one.

Example 2. Show that the set $\mathscr{I}$ of all integers is countable.

Solution. A simple arrangement of the integers in a sequence is:

$$(1) \qquad\qquad 0, 1, -1, 2, -2, 3, -3, \cdots.$$

A formula for the nth integer a_n is: If n is even, $a_n = \frac{1}{2}n$. If n is odd, $a_n = \frac{1}{2}(1 - n)$. This correspondence is clearly one-to-one, and every integer appears in the sequence $\{a_n\}$.

1213. COUNTABILITY OF THE RATIONAL NUMBERS

Since between any two natural numbers there are infinitely many rational numbers, it may seem almost paradoxical that there can exist a one-to-one correspondence between the natural numbers and the rational numbers. Such a correspondence is possible, however, as we shall demonstrate in this section. We do this in two steps.

Theorem I. *The set A of positive rational numbers is countably infinite.*

Proof. We shall place the positive rational numbers in an infinite sequence by first arranging them in a doubly infinite array as follows:

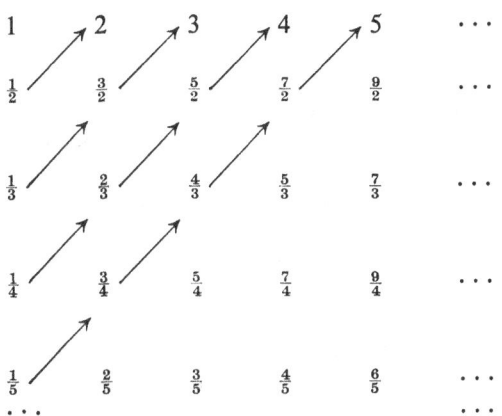

Following the path indicated by the arrows we obtain the desired sequence:

$$1, \tfrac{1}{2}, 2, \tfrac{1}{3}, \tfrac{3}{2}, 3, \tfrac{1}{4}, \tfrac{2}{3}, \tfrac{5}{2}, 4, \tfrac{1}{5}, \tfrac{3}{4}, \cdots .$$

We designate these positive rational numbers by the symbols $\{p_n\}$: $p_1 = 1$, $p_2 = \tfrac{1}{2}, p_3 = 2, \cdots$. It is possible to obtain an explicit formula for p_n as a function of the natural number n, but such a formula is far too cumbersome to be worthwhile for our purposes. It is sufficient to know that a rigorous procedure is established to determine each particular p_n.

Theorem II. *The set $\mathscr{2}$ of all rational numbers is countably infinite.*

Proof. By following the ideas of Example 2, § 1212, and using the sequence $\{p_n\}$ of positive rational numbers from Theorem I, we have the entire set of rational numbers arranged in a sequence:

$$0, p_1, -p_1, p_2, -p_2, p_3, -p_3, p_4, -p_4, \cdots .$$

NOTE. Let Ω be any nonempty family of open intervals I. The density of the set $\mathscr{2}$ of rational numbers guarantees that for each I of Ω the set $A(I) \equiv \{r \mid r \in \mathscr{2},\ r \in I\}$ of all rational numbers in I is nonempty. The countability of $\mathscr{2}$ provides a rule that selects a member $r = r(I)$ of $A(I)$ for every I of Ω as follows: Let $\{r_n\}$ be an arrangement of the members of $\mathscr{2}$ as a sequence, and let $k = k(I)$ be the least subscript n such that $r_n \in A(I)$; that is, $r_k \in A(I)$ and for no m less than k does the statement $r_m \in A(I)$ hold. Then $r = r(I)$ is defined to be r_k. The significance of this fact is that a single rule of selection has been set up to apply to the entire family Ω. If Ω were an arbitrary nonempty family of nonempty sets S, then a rule of selection that chooses a member from each set S of the collection Ω would exist only by means of an added postulate that such a *choice function* does indeed exist. This postulate is known as the **axiom of choice.** For a discussion of the axiom of choice and other equivalent formulations, see E. J. McShane and T. A. Botts, *Real Analysis* (Princeton, D. Van Nostrand Co., Inc., 1959), Appendix II. One of the alternative formulations of the axiom of choice is the principle of transfinite induction mentioned in the footnote, § 303. It is for the reasons outlined above that the existence of the rational numbers r_n and s_n for every positive integer n, as used in the proofs of Theorem II, § 1102, and the Theorem, § 1104, can be assumed without use of the axiom of choice.

A simple illustration of the use to which the choice function $r(I)$ described in the preceding paragraph can be put is a proof of the following fact (a corresponding fact concerning strictly decreasing sequences of rational numbers also holds): *If x is any real number, then there exists a strictly increasing sequence of rational numbers whose limit is x.* We construct the strictly increasing sequence $\{s_n\}$ of rational numbers as follows: Let $\{r_n\}$ be an arrangement of the members of $\mathscr{2}$ as a sequence, and define s_1 to be the first term of the sequence $\{r_n\}$ less than x: $s_1 < x$. The second term s_2 is then defined to be the first term of the sequence $\{r_n\}$ between s_1 and x: $s_1 < s_2 < x$. In general, if n is an arbitrary positive integer for which s_n is defined, s_{n+1} is defined to be the first term of the sequence $\{r_n\}$ between s_n and x: $s_n < s_{n+1} < x$. In this way a strictly increasing sequence $\{s_n\}$ of rational numbers all of which are less than x is inductively defined. Since x is an upper bound of $\{s_n\}$, in order to show that

$\lim_{n \to +\infty} s_n = x$, or sup $\{s_n\} = x$, it is necessary only to show that $\{s_n\}$ has no upper bound less than x. Let us assume that there *does* exist such an upper bound y less than x: $s_n \leqq y < x$ for $n = 1, 2, \cdots$. We can obtain a contradiction as follows: In the first place, if for each positive integer n, m_n is the subscript for the term of the sequence r_n that gives the nth term of the sequence $\{s_n\}$, that is, $s_n = r_{m_n}$, then (by the definitions of s_n and s_{n+1}) $m_{n+1} > m_n$ for $n = 1, 2, \cdots$. Therefore, by Example 4, § 1204, $m_n \geqq n$ for $n = 1, 2, \cdots$. Now let r_N be the first term of the sequence $\{r_n\}$ such that $y < r_N < x$. By definition of s_{n+1} and by the fact that for every n, $s_n \leqq y < x$, we know that $m_{n+1} \leqq N$. We conclude that $N \geqq n + 1$ for every positive integer n. The desired contradiction is given by the Archimedean property (§ 903).

1214. UNCOUNTABILITY OF THE REAL NUMBERS

We shall now prove that it is impossible to arrange the real numbers in a sequence—there are simply too many of them. The procedure will be to show that the real numbers x such that $0 \leq x < 1$ are too numerous to be arranged in a sequence, and it will follow that the larger set $\mathscr{R}$ of *all* real numbers is too numerous. (The idea here is that if *all* real numbers could be arranged $x_1, x_2, x_3, \cdots$, then we could define a sequence $y_1, y_2, y_3 \cdots$ of the points of $[0, 1)$ by letting y_1 be the first x_n that belongs to $[0, 1)$, by letting y_2 be the next x_n that belongs to $[0, 1)$, and so forth inductively.)

By Theorem V, §1207, the points of $[0, 1)$ are in a one-to-one correspondence with all sequences of digits $0, 1, 2, \cdots, 9$ such that infinitely many terms are different from 9. If we assume—as we now do—that the points of $[0, 1)$ can be arranged in a sequence, then we are assuming at the same time that all sequences of digits of the type described above—which we shall call **canonical sequences**—can be arranged in a sequence. We shall obtain a contradiction to our assumption by showing that *any arbitrary* sequence of canonical sequences must *inevitably* omit at least one canonical sequence. In other words, no matter how we try to arrange the canonical sequences in a sequence, there will always be at least one left over.

To designate a sequence of sequences we use double subscripts. The following double array, then, represents our assumed arrangement of all canonical sequences in a sequence of sequences; the first (horizontal) row represents the first canonical sequence, the second row represents the second, and so forth:

$$(1) \quad \begin{bmatrix} d_{11}, & d_{12}, & d_{13}, & d_{14}, & \cdots, & d_{1n}, & \cdots \\ d_{21}, & d_{22}, & d_{23}, & d_{24}, & \cdots, & d_{2n}, & \cdots \\ d_{31}, & d_{32}, & d_{33}, & d_{34}, & \cdots, & d_{3n}, & \cdots \\ d_{41}, & d_{42}, & d_{43}, & d_{44}, & \cdots, & d_{4n}, & \cdots \\ \cdots & & & & & & \cdots \\ d_{m1}, & d_{m2}, & d_{m3}, & d_{m4}, & \cdots, & d_{mn}, & \cdots \\ \cdots & & & & & & \cdots \end{bmatrix}$$

The first subscript tells which canonical sequence is designated, and the second subscript designates the term of the canonical sequence specified by the first subscript.

We now set to work to find a canonical sequence that is not in the array given above. We do this by what is called the *diagonal process*. This consists of constructing a sequence of digits $e_1, e_2, e_3, \cdots$ such that $e_1 \neq d_{11}$, $e_2 \neq d_{22}$, $e_3 \neq d_{33}$, and, in general, $e_n \neq d_{nn}$ for every natural number n. It is also important that this new sequence $\{e_n\}$ be canonical—that is, that infinitely many of the terms be different from 9. This part is easy—we shall see to it that *all* terms are different from 9. In fact, all we need do is to pick any two distinct digits, say 1 and 2, and assign these as values of e_n in such a way that e_n is always different from d_{nn}. A simple way of doing this is to let e_n be defined to be 1 whenever d_{nn} is even, and to let e_n be defined to be 2 whenever d_{nn} is odd. Then, since e_n is odd whenever d_{nn} is even, and even whenever d_{nn} is odd, the equation $e_n = d_{nn}$ can never hold. We can now draw our conclusion: the sequence $\{e_n\}$ is not in the array (1): it is not the first sequence since its first term is different; it is not the second sequence since its second term is different; in general, for every natural number n, it is not the nth sequence since its nth term is different. The desired contradiction has been reached, and the proof is complete.

Example. If the double array (1) is as follows:

(2)
$$\begin{cases} 1, & 7, & 1, & 7, & 1, & 7, & \cdots \\ 5, & 0, & 0, & 0, & 0, & 0, & \cdots \\ 2, & 2, & 2, & 2, & 2, & 2, & \cdots, \\ 5, & 5, & 5, & 5, & 5, & 5, & \cdots \\ 2, & 4, & 7, & 2, & 4, & 7, & \cdots \\ 6, & 1, & 6, & 1, & 6, & 1, & \cdots \end{cases}$$

then the sequence $\{e_n\}$ is

$$2, 1, 1, 2, 1, 2, \cdots.$$

We summarize our results in a formal statement, and derive a corollary:

Theorem I. *The set $\mathscr{R}$ of real numbers is uncountable.*

Theorem II. *The set A of irrational numbers is uncountable.*

Proof. Assume that A is countable, and let its members be arranged in a sequence: $a_1, a_2, \cdots, a_n, \cdots$. If $r_1, r_2, \cdots, r_n, \cdots$ is the sequence of all rational numbers, then the real numbers can be arranged in a sequence as follows:

$$r_1, a_1, r_2, a_2, r_3, a_3, r_4, a_4, \cdots.$$

This contradiction to Theorem I is the one desired.

NOTE. An **algebraic real number** is a real number that is a root of a nonzero polynomial all of whose coefficients are integers. A **transcendental number** is a real number that is not algebraic.   It can be shown (cf. the Birkhoff and MacLane text cited in the footnote of § 606) that the algebraic numbers constitute an ordered field, and that they form a countable set.   It follows from the uncountability of the real number system that the set of transcendental numbers is also uncountable.   (Why?) Consequently, transcendental numbers must exist.   The most familiar transcendental numbers are e (the base of the so-called system of natural logarithms) and π.   The transcendental character of e was established by C. Hermite (1822–1905, French) in 1873, and that of π by C. L. F. Lindemann (1852–1939, German) in 1882.†   The first number to be proved transcendental was neither e not π, but a number constructed artificially for the purpose by J. Liouville (1809–1882, French) in 1844 (cf. page 413 of the Birkhoff and MacLane reference).   The demonstration of the uncountability of the transcendental numbers is due to G. Cantor (1845–1918, German).

† It is easy to show that e is irrational (for example, cf. Ex. 34, § 1211, of the author's *Advanced Calculus*, in the Appleton-Century Mathematics Series).   A proof of the irrationality of π was given in 1761 by J. H. Lambert (1728–1777, German).   For an elementary proof, and further discussion of both the irrationality and transcendence of π, see Ivan Niven, *Irrational Numbers* (Carus Mathematical Monographs, 1956).

Appendix

....

I. INTRODUCTION

The purpose of this appendix is to present several alternative formulations of the completeness property for the real number system and to establish equivalences among them. Although all necessary definitions and theorems are stated, standard proofs that are to be found in any fairly complete textbook on mathematical analysis at the level of advanced calculus—for example, the author's *Advanced Calculus* in the Appleton-Century Mathematics Series—are omitted. No effort at motivation or complete discussion will be made, and the general demands on the reader by way of maturity, independence, and experience will be greater than in the main text of the book.

In the present section we give the definitions and theorems needed in the sequel, with references to the author's *Advanced Calculus* (abbreviated *AC*) when proofs are omitted. In §2 we give ten statements each of which is equivalent, in an ordered field, to the Axiom of Completeness of § 902, and we establish their equivalences by a cycle of implications. In § 3 we give four more methods of imposing completeness, each involving a property of continuous functions on a closed interval. Finally, in § 5, we present an additional triplet of properties each of which is equivalent to completeness in an *Archimedean* ordered field. Seven more formulations of completeness in terms of continuity are given in Exercises 7–13, § 6. References to the literature of non-Archimedean ordered fields are given at the end of § 713. For further comments and historical sidelights, see E. T. Bell, *The Development of Mathematics* (New York, McGraw Hill Book Co., 1940), especially p. 439.

Definition I. *A nonempty set S with a relation < (cf. § 309) is a **totally ordered system**, or a **simply ordered system**, or a **chain** if and only if (i) the statement a < a is false for every a ∈ S; (ii) transitivity holds: a < b and b < c imply a < c; (iii) trichotomy holds: for any a and b in S, either a < b, a = b, or b < a. The relation > is defined by the equivalence of a < b and b > a.*

Theorem I. *In a totally ordered system S no two of the three statements a < b, a = b, and a > b can hold simultaneously.*

Proof. This is a consequence of (*i*) and (*ii*).

Theorem II.   *An ordered field is a totally ordered system.*

Proof. This was proved in Chapter 2.

Definition II.   *Let S be a totally ordered system.* **Open** *and* **closed intervals** *in S are defined by inequalities as in § 802.   A subset A of S is* **open** *if and only if corresponding to each point x of A there exists an open interval of S that contains x and is a subset of A.   A point x is a* **limit point** *of a subset A of S if and only if every open interval of S that contains x contains at least one point of A different from x.   A subset A of S is* **closed** *if and only if every limit point of A is a member of A.   If A is any subset of S, the* **complement** *A' of A is the set of all points of S that are not members of A.*

Definition III.   *Two sets A and B in a totally ordered system S are* **separated** *if and only if* (*i*) *A and B are both nonempty,* (*ii*) *A and B have no point in common, and* (*iii*) *no point of A is a limit of B and no point of B is a limit point of A.   A subset C of S is* **connected** *if and only if there do not exist separated sets A and B such that* (*i*) *A and B are subsets of C and* (*ii*) *every member of C is either a member of A or a member of B.*

Theorem III.   *In any totally ordered system S the empty set ∅ (§ 201) is both open and closed.*

Proof. If ∅ were not open it would contain a point that belongs to no open interval of *S* lying in ∅, and hence ∅ would contain a point.   The empty set has no limit points and therefore contains all of its limit points.

Theorem IV.   *In any totally ordered system S a set is open if and only if its complement is closed; equivalently, a set is closed if and only if its complement is open.*

Proof. If *A* is open and *A'* not closed then there exists a point *x* of *A* that is a limit point of *A'*.   But there is an open interval of *S* containing *x* and lying entirely in *A*, so that *x* is *not* a limit point of *A'*.   If *A'* is closed and *A* not open then there exists a point *x* of *A* such that *no* open interval of *S* containing *x* lies entirely in *A;* that is, *every* open interval of *S* that contains *x* must contain at least one point of *A'*, and *x* is a limit point but not a member of *A'*, and *A'* is not closed.

Theorem V.   *If x is a limit point of a set A in a totally ordered system S, then every open interval of S containing x must contain infinitely many points of A.*

Proof.   Assume that *x* is a limit point of *A*, that $a < x < b$, and that there are only finitely many points $a_1, a_2, \cdots, a_n$ of *A* that lie between *a* and *b*.   This means that if any of these points lie between *a* and *x* there must be only finitely many and therefore, by the methods used in the solution of Example

4, § 306, there must be a greatest. Let c denote this greatest member of the finite set less than x—unless there are none, in which case let $c \equiv a$. Similarly, if there are any of $\{a_1, a_2, \cdots, a_n\}$ between x and b there will be a least member. Denote by d this least member—or, if there are none, let $d \equiv b$. As a consequence of this we have an open interval (c, d) containing x but no *other* point of A, and x is *not* a limit point of A. (Contradiction.)

Definition IV. *If A_α, for each α of an index set (that is, a set of indices or subscripts) Σ, is a set of points in a totally ordered system S the **union** $\underset{\alpha}{\cup} A_\alpha$ of the sets A_α consists of points p such that $p \in A_\alpha$ for at least one $\alpha \in \Sigma$. The **intersection** $\cap A_\alpha$ of the sets A_α consists of points p such that $p \in A_\alpha$ for every $\alpha \in \Sigma$. If $B \subset \underset{\alpha}{\cup} A_\alpha$, then the collection $\{A_\alpha\}$ of the sets A_α is called a **covering** of B. If in addition every set A_α is open the covering $\{A_\alpha\}$ is called an **open covering**.*

The definitions of increasing, decreasing, and monotonic sequences of members of an ordered field, given in §§ 316, 1003, and 1204, are immediately available without change for any totally ordered system.

Definition V. *A sequence $\{b_n\}$ is a **subsequence** of a sequence $\{a_n\}$ if and only if there exists a strictly increasing sequence $\{m_n\} = m_1, m_2, m_3, \cdots, m_n, \cdots$ of natural numbers such that $b_n = a_{m_n}$ for every natural number n.*

Theorem VI. *If a sequence in a totally ordered system is increasing, strictly increasing, decreasing, strictly decreasing, monotonic, strictly monotonic, bounded above, bounded below, or bounded, so is any subsequence.*

Proof. Trivial.

Theorem VII. *Every sequence in a totally ordered system contains a monotonic subsequence.*

Proof. The essential details of the proof are contained in the proof of Lemma 1, § 217, of *AC*.

Definition VI.† *A sequence $\{a_n\}$ in an ordered field $\mathscr{G}$ **converges** if and only if there exists a member a of $\mathscr{G}$ having the property: corresponding to an arbitrary positive member ϵ of $\mathscr{G}$ there exists a natural number N such that the inequality $n > N$ implies $|a_n - a| < \epsilon$. The element a is called the **limit** of the sequence $\{a_n\}$, with the notation*

$$\lim_{n \to +\infty} a_n = a, \text{ or } a_n \to a \text{ as } n \to +\infty$$

*and the sequence is said to **converge to** a.* (Cf. the Theorem, § 1204.)

---

† Most of the concepts and facts in the remainder of this section can be extended to totally ordered systems in general. However, for the sake of simplicity, since our interest is principally focused on ordered fields, we have formulated statements in those terms.

Theorem VIII. *Every convergent sequence in an ordered field is bounded and has a unique limit.*

Proof. Assume $a_n \to a$ as $n \to +\infty$, and let $\epsilon \equiv 1 > 0$. Then there exists a natural number N such that $n > N$ implies $|a_n - a| < 1$, or $a - 1 < a_n < a + 1$. Therefore, for $n = 1, 2, \cdots$, min $(a - 1, a_1, a_2, \cdots, a_N) \leqq a_n \leqq$ max $(a + 1, a_1, a_2, \cdots, a_N)$, and $\{a_n\}$ is bounded. Assume $a_n \to a$ and $a_n \to b$ as $n \to +\infty$, where $a < b$, and let $\epsilon \equiv \frac{1}{2}(b - a)$. Then there exists a natural number N_1 such that $n > N_1$ implies $|a_n - a| < \epsilon$, and there exists a natural number N_2 such that $n > N_2$ implies $|a_n - b| < \epsilon$. If $n \equiv 1 + \max(N_1, N_2)$, then $b - \epsilon < a_n < a + \epsilon$, and hence $b - a < 2\epsilon = b - a$. (Contradiction.)

Theorem IX. *In any ordered field, every subsequence of a convergent sequence is convergent, and its limit is the same as the limit of the original sequence.*

Proof. If $\{b_n\}$ is a subsequence of $\{a_n\}$, with $b_n = a_{m_n}$, if $a_n \to a$, and if $\epsilon > 0$, let N be such that $n > N$ implies $|a_n - a| < \epsilon$. Then $n > N$ implies (cf. Example 4, § 1204) $m_n \geqq n > N$, and hence $|b_n - a| = |a_{m_n} - a| < \epsilon$.

Theorem X. *In any ordered field $\mathcal{G}$ an increasing sequence $\{a_n\}$ converges if and only if the set of its values has a supremum λ; in case of convergence, $a_n \to \lambda$ as $n \to +\infty$. A decreasing sequence $\{b_n\}$ converges if and only if the set of its values has an infimum γ; in case of convergence, $b_n \to \gamma$ as $n \to +\infty$. Any monotonic sequence that contains a convergent subsequence is convergent. The set of values of a monotonic sequence has at most one limit point. A strictly monotonic sequence converges if and only if the set of its values has a limit point.*

Proof. If $\{a_n\}$ is an increasing sequence converging to a, then a is an upper bound for $\{a_n\}$ since otherwise there would exist a natural number N such that $a_N > a$. Defining $\epsilon \equiv a_N - a$, we have, for every $n > N$, $a_n \geqq a_N$ and hence $|a_n - a| = a_n - a \geqq a_N - a = \epsilon > 0$, in contradiction to $a_n \to a$. Furthermore, if a is not the *least* upper bound of $\{a_n\}$ there exists a member b of $\mathcal{G}$ such that for every $n = 1, 2, \cdots, a_n \leqq b < a$. If $\epsilon \equiv a - b$, then the inequality $|a_n - a| < \epsilon$ can never hold, in contradiction to $a_n \to a$. Conversely, assume that λ is the supremum of the set of values of the increasing sequence $\{a_n\}$, and let $\epsilon > 0$. Then there exists a natural number N such that $a_N > \lambda - \epsilon$, and therefore $n > N$ implies $\lambda - \epsilon < a_N \leqq a_n \leqq \lambda < \lambda + \epsilon$, and hence $|a_n - \lambda| < \epsilon$. Therefore $a_n \to \lambda$ as $n \to +\infty$. The corresponding details for a decreasing sequence are completely analogous, with a reversal of inequalities. Now let $\{a_n\}$ be a monotonic sequence containing the convergent subsequence $\{b_n\}$, and assume for definiteness that $\{a_n\}$ is increasing. From what has just been established, we have only to prove that $\sigma \equiv \sup \{b_n\}$ is the least upper bound of $\{a_n\}$. In the first place, σ is an upper bound of $\{a_n\}$ since otherwise there would exist an n such that

$a_n > \sigma$, and hence $b_n = a_{m_n} \geqq a_n > \sigma$. On the other hand there can be no upper bound b of $\{a_n\}$ *less* than σ, since such an upper bound of $\{a_n\}$ would also be an upper bound of $\{b_n\}$ less than the least upper bound. Next, we assume there exists a monotonic sequence $\{a_n\}$ whose set of values has at least two distinct limit points b and c, assuming further (for definiteness) that $\{a_n\}$ is increasing and that $b < c$. A contradiction is obtained by considering the open interval $I = (b, 2c - b)$ containing c. Since c is a limit point of $\{a_n\}$ there must exist a natural number N such that $a_N \in I$, and therefore $n > N$ implies $a_n \geqq a_N > b$. This means that the open interval $(2b - a_N, a_N)$ containing b contains at most the first $N - 1$ terms of the sequence $\{a_n\}$. Therefore, by Theorem V, b cannot be a limit point of $\{a_n\}$. Finally, let $\{a_n\}$ be a strictly increasing sequence, and assume first that $a_n \to a$ as $n \to +\infty$. Then, as has already been proved, a is the least upper bound of $\{a_n\}$. Furthermore, since $\{a_n\}$ has no largest term, the inequality $a_n < a$ must hold for all n. Therefore every open interval containing a must contain at least one term of the sequence different from a, and a is a limit point of $\{a_n\}$. Conversely, assume that b is a limit point of $\{a_n\}$. Then b must be an upper bound for $\{a_n\}$, for otherwise, if N is such that $a_N > b$, then the open interval $(2b - a_N, a_N)$ containing b contains only finitely many points of $\{a_n\}$, and b is not a limit point of $\{a_n\}$. The limit point b must be the *least* upper bound of $\{a_n\}$, for otherwise, if c is an upper bound of $\{a_n\}$ and $c < b$, since $a_n < c$ for all n, the open interval $(c, 2b - c)$ containing b contains no points of $\{a_n\}$, and b is not a limit point of $\{a_n\}$. Since the final details for decreasing sequences are strict duplicates of those just presented for increasing sequences, the proof is complete.

2. ALTERNATIVES FOR COMPLETENESS

Theorem. *Let $\mathscr{G}$ be an ordered field. Then each of the following is a necessary and sufficient condition for $\mathscr{G}$ to be complete:*

   I. *Every nonempty set of members of $\mathscr{G}$ that is bounded below has a greatest lower bound.*

   II. *The **Bolzano–Weierstrass**† property holds: Every infinite bounded set has at least one limit point.*

   III. *$\mathscr{G}$ is connected.*

   IV. *The **overlapping intervals property**‡ holds: If $\{[a_\alpha, b_\alpha]\}$ is a nonempty family of closed intervals every pair of which have at least one point in common then there exists at least one point x belonging to every interval of the family: $x \in \bigcap_\alpha [a_\alpha, b_\alpha]$.*

† After B. Bolzano (1781–1848, of Prague, then Austrian) and K. W. T. Weierstrass (1815–1897, German).

‡ Due to G. Cantor (1845–1918, German). Also cf. part (*ii*) of Theorem IV, § 4, for the related *nested intervals property*. The extension to general topological spaces is known as the *finite intersection property*. Cf. John L. Kelley, *General Topology* (New York, D. Van Nostrand Co., Inc., 1955).

V. **The Dedekind property**† *holds: If $\mathscr{G}$ is divided into two nonempty sets L and R whose union is $\mathscr{G}$ and which are such that if x is an arbitrary member of L and y is an arbitrary member of R then $x < y$, then either L has a greatest member or R has a least member.*

VI. **The Heine–Borel property** § *holds: If $\{A_\alpha\}$ is an open covering of a closed interval $[a, b]$, then there exists a finite subcollection of $\{A_\alpha\}$, $A_{\alpha_1}, A_{\alpha_2}, \cdots, A_{\alpha_n}$, that is a covering of $[a, b]$; in short, every open covering of a closed interval can be reduced to a finite covering.*

VII. *Every bounded increasing sequence converges.*

VIII. *Every bounded decreasing sequence converges.*

IX. *Every bounded monotonic sequence converges.*

X. *Every bounded sequence contains a convergent subsequence.*

Proof. The technique of proof will be to establish a complete cycle of implications: *The axiom of completeness* (§ 902) *implies I, I implies II, II implies III, $\cdots$, X implies the axiom of completeness,*

The axiom of completeness implies I: This is the Theorem, § 902.

I implies II: Assume *I* and that there exists an infinite bounded set A without a limit point. Let a be a lower bound of A and b be an upper bound of A so that $z \in A$ implies $a \leq z \leq b$, and let B be the set of all x of the closed interval $[a, b]$ such that the set $\{y \mid y \in A, x \leq y \leq b\}$ is finite. Then $b \in B$, so that B is a nonempty set bounded below (by a). Let γ be the greatest lower bound of B: $\gamma \equiv \inf B$. We shall show that γ is a limit point of A by assuming the contrary, that is, that there is an open interval $I = (c, d)$ containing γ but no point of A different from γ. If $a < \gamma < b$ then there is a point u of the interval I between a and γ, and there is a point x of the interval I between γ and b. But since there are only finitely many points y of A such that $x \leq y \leq b$ and at most one point y of A such that $u \leq y < x$ (namely, $y = \gamma$), there are only finitely many points y of A such that $u \leq y \leq b$, and hence $u \in B$. But u is less than $\inf B$. (Contradiction.) If $\gamma = b$, then there is a point u of the interval I between a and b, and since there is at most one point y of A such that $u \leq y \leq b$ (namely, $y = \gamma = b$), $u \in B$. But u is less than $\inf B$. (Contradiction.) Therefore $\gamma = a$, and there is a point x of the interval I between $a = \gamma$ and b. But since there are only finitely many points y of A such that $x \leq y \leq b$ and at most one point y of A such that $a = \gamma \leq y < x$ (namely, $y = a = \gamma$), there are only finitely many points y of A such that $a \leq y \leq b$. But all points y of the infinite set A satisfy these inequalities. (Contradiction.)

II implies III: We first establish a lemma:

Lemma. *If II is true, and if $p > 0$, then $p/n \to 0$ and $p/2^n \to 0$ as $n \to +\infty$.*

Proof of lemma. Since $\{p/n\}$ is a strictly decreasing sequence bounded below by 0 and above by p, and since its set of values is assumed to have a

† After J. W. R. Dedekind (1831–1916, German).
§ After E. Heine (1821–1881, German) and E. Borel (1871–1956, French).

limit point, the sequence must converge to the greatest lower bound of its values, by Theorem X, § 1. Since 0 is a lower bound, if $\{p/n\}$ does not converge to 0 it must converge to a positive limit: $p/n \to c > 0$ as $n \to +\infty$, where $c = \inf \{p/n\}$, and $c < p/n$ for every $n = 1, 2, \cdots$. In particular (for $n = 1$), $c < p$, and hence $c < cp/(p - c)$. Therefore there exists a natural number N such that $n > N$ implies $p/n < cp/(p - c)$ or, equivalently, $p/(n + 1) < c$. (Contradiction.) Finally, since $\{p/2^n\}$ is a subsequence of $\{p/n\}$, $p/2^n \to 0$ as $n \to +\infty$.

II implies III, continued: Assume that II is true and that $\mathscr{G}$ is not connected. Then $\mathscr{G}$ can be represented as the union of two separated sets A and B. Let a_1 be a member of A and b_1 be a member of B, and assume for definiteness that $a_1 < b_1$. Consider the midpoint $\frac{1}{2}(a_1 + b_1)$ of the closed interval $[a_1, b_1]$. If this midpoint belongs to A call it a_2, define $b_2 \equiv b_1$, and consider the midpoint $\frac{1}{2}(a_2 + b_2)$ of the right-hand half-interval $[a_2, b_2]$. If the midpoint of $[a_1, b_1]$ belongs to B call it b_2, define $a_2 \equiv a_1$, and consider the midpoint $\frac{1}{2}(a_2 + b_2)$ of the left-hand half-interval $[a_2, b_2]$. In general, if $\frac{1}{2}(a_n + b_n) \in A$, call it a_{n+1} and define $b_{n+1} \equiv b_n$; if $\frac{1}{2}(a_n + b_n) \in B$, call it b_{n+1} and define $a_{n+1} \equiv a_n$. In this way, two bounded monotonic sequences are obtained inductively, $\{a_n\}$ increasing and $\{b_n\}$ decreasing, with $a_n < b_n$, and $b_n - a_n = (b_1 - a_1)/2^{n-1}$. If there exists a natural number N such that $n > N$ implies $a_n = a_N$, then $a_n \to a_N$ as $n \to +\infty$ and $\{a_n\}$ is a convergent sequence. On the other hand, if there exists no such N, define $m_1 \equiv 1$ and for any n define m_{n+1} to be the least natural number k greater than m_n such that $a_k > a_{m_n}$. If $c_n \equiv a_{m_n}$, then $\{c_n\}$ is a *strictly* increasing subsequence of $\{a_n\}$. Since $\{c_n\}$ is bounded it converges by assumption II and Theorem X, § 1, and furthermore (by the same theorem) $\{a_n\}$ converges also. Under all circumstances, then, $\{a_n\}$ converges. By similar reasoning we conclude that $\{b_n\}$ converges too. Assume $a_n \to a$ and $b_n \to b$. Since, for any natural numbers m and n, if $p \equiv \max{(m, n)}$, $a_m \leq a_p < b_p \leq b_n$, every term of $\{a_n\}$ is a lower bound for $\{b_n\}$, and hence for every n, $a_n \leq b$. Consequently $a \leq b$. If $a < b$, since $a_n \leq a$ for all n and since, by the preceding lemma, $b_n - a_n \to 0$, it follows that for sufficiently large n, $b_n - a_n < b - a$, or $b_n < a_n + b - a \leq b$. But this contradicts the inequality $b_n \geq b$, and we infer that $a = b$; that is, that $\{a_n\}$ and $\{b_n\}$ have the same limit $a = b$. Finally, we seek a contradiction to the assumption that this limit is a member of the set A (if the limit is a member of B, a contradiction can be obtained in an identical manner). Accordingly, we have $b_n \to a$, where $a \in A$, and $b_n \in B$ for every natural number n. By assumption, a is not a limit point of B and therefore there exists an open interval (u, v) containing a but no point of B. If we define ϵ to be the positive number $\epsilon \equiv \min{(a - u, v - a)}$, then the inequality $|b_n - a| < \epsilon$ is never satisfied for any b_n. This contradiction to the fact that $b_n \to a$ is the one sought.

III implies IV: Assume that $\mathscr{G}$ is connected and that there exists a nonempty family $\{[a_\alpha, b_\alpha]\}$ of closed intervals each pair of which have a point in common, but whose intersection $\bigcap_\alpha [a_\alpha, b_\alpha]$ is empty.   Let A be the set of all points x of $\mathscr{G}$ such that $x < a_\alpha$ for *some* α, and let B be the set of all points y of $\mathscr{G}$ such that $y > b_\alpha$ for *some* α.   Then $\mathscr{G}$ is equal to the union of A and B, for otherwise there would exist a point w belonging to neither A nor B and hence such that $w \geq a_\alpha$ for every α and $w \leq b_\alpha$ for every α.   Therefore $w \in [a_\alpha, b_\alpha]$ for every α.   (Contradiction.)   Since points a_α and b_α exist, A and B are nonempty.   Finally A and B have no point in common, for if u were in *both* A and B, there would exist α and β such that $u < a_\alpha$ and $u > b_\beta$.   But this would mean that the closed intervals $[a_\alpha, b_\alpha]$ and $[a_\beta, b_\beta]$ would have no point in common.   (Contradiction.)   We conclude that either a point a of A is a limit point of B or a point b of B is a limit point of A, since $\mathscr{G}$ is connected.   Assume for definiteness that there exists a point a of A that is a limit point of B.   (The reader can supply the details for the alternative.) Since $a < a_{\alpha_0}$ for some α_0 and since $a_{\alpha_0} \leq b_\alpha$ for *every* α (otherwise the closed intervals $[a_\alpha, b_\alpha]$ and $[a_{\alpha_0}, b_{\alpha_0}]$ would have no point in common), the open interval $(2a - a_{\alpha_0}, a_{\alpha_0})$ contains a and no point of B.   (Contradiction.)

IV implies V: Assume IV and that the Dedekind Property V fails; that is, that there exist two nonempty sets L and R whose union is the ordered field $\mathscr{G}$, such that $x \in L$ and $y \in R$ imply $x < y$, and such that L has no greatest member and R has no least member.   Consider the family of all closed intervals $[x, y]$, where $x \in L$ and $y \in R$.   Any two of these, $[x_1, y_1]$ and $[x_2, y_2]$, have a point in common, since any point u satisfying the inequalities $\max(x_1, x_2) \leq u \leq \min(y_1, y_2)$ belongs to both.   Therefore, by IV, there exists a point c belonging to every interval $[x, y]$; that is, $x \in L$ and $y \in R$ imply $x \leq c \leq y$.   Since $c \in \mathscr{G}$, either $c \in L$ or $c \in R$.   If $c \in L$, then c must be the greatest member of L, and if $c \in R$, then c must be the least member of R.   (Contradiction.)

V implies VI: Assume V and that the Heine-Borel property fails for the closed interval $[a, b]$, and let $\{A_\alpha\}$ be an open covering of $[a, b]$ that cannot be reduced to a finite covering.   Define a set L to consist of all members x of the ordered field $\mathscr{G}$ such that $x \leq a$, together with all those x (if any) between a and b such that the closed interval $[a, x]$ can be covered by a finite subcollection of $\{A_\alpha\}$.   Define a set R to consist of all members x of $\mathscr{G}$ such that $x > b$, together with all those x such that $a < x \leq b$ and such that the closed interval $[a, x]$ *cannot* be covered by a finite subcollection of $\{A_\alpha\}$.   Clearly, L and R are nonempty sets whose union is $\mathscr{G}$, and $x \in L$ and $y \in R$ imply $x < y$.   Assume first that L has a greatest member c.   Then $a \leq c < b$, and the point c must belong to some set A_γ of the covering.   Since A_γ is open it contains an open interval I containing c, and there must be a point v of I between c and b.   If $c = a$, then $[a, v]$ is finitely covered by A_γ, and $v \in L$. If $c > a$, and if $\{A_{\alpha_1}, A_{\alpha_2}, \cdots, A_{\alpha_n}\}$ is a finite covering of $[a, c]$, then $\{A_{\alpha_1}, A_{\alpha_2}, \cdots, A_{\alpha_n}, A_\gamma\}$ is a finite covering of $[a, v]$, and $v \in L$.   But in either case, $v > c$ and a contradiction has been obtained to the assumption that c is the greatest member of L. Finally, we assume that R has a least member c. Then $a < c \leqq b$, and the point c must belong to some set A_γ of the covering. Since A_γ is open it contains an open interval I containing c, and there must be a point u of I between a and c. Since $u < c, u \in L$ and therefore there exists a finite covering $\{A_{\alpha_1}, A_{\alpha_2}, \cdots, A_{\alpha_n}\}$ of the closed interval $[a, u]$. Then $\{A_{\alpha_1}, A_{\alpha_2}, \cdots, A_{\alpha_n}, A_\gamma\}$ is a finite covering of $[a, c]$. (Contradiction.)

VI implies VII: Assume VI and that there exists a bounded increasing sequence $\{a_n\}$ that fails to converge, and assume that $a \leqq a_n \leqq b$ for every n. As shown in the proof of II implies III, there must exist a strictly increasing subsequence $\{c_n\}$ of $\{a_n\}$. By Theorem X, § 1, this sequence fails to converge, and the set of its values has no limit point. Therefore, the set $C_1 \equiv \{c_1, c_2, \cdots, c_n, \cdots\}$ is closed. Similarly, the set $C_2 \equiv \{c_2, c_3, \cdots, c_n, \cdots\}$, being a subset of C_1, has no limit point and is therefore closed. In general, the set $C_n \equiv \{c_n, c_{n+1}, \cdots\}$ is closed for every $n = 1, 2, \cdots$. If G_n is defined to be the complement (in the ordered field $\mathscr{G}$) of the set C_n, for every $n = 1, 2, \cdots$: $G_n \equiv C_n'$, then by Theorem IV, § 1, G_n is open for $n = 1, 2, \cdots$. The collection $\{G_n\}$ is an open covering of the closed interval $[a, b]$ since (for every n), $c_n \in G_{n+1}$, and any point of $[a, b]$ that is not in the sequence $\{c_n\}$ belongs to every G_n. Therefore, by assumption, the interval $[a, b]$ is a subset of the union of a finite number of the G_N's and hence, since $G_n \subset G_{n+1}$ for every n, $[a, b]$ is a subset of G_n for some n (the largest of the finite set of subscripts). But this means that c_n is a member of G_n, whereas $c_n \in C_n$. (Contradiction).

VII implies VIII: If $\{a_n\}$ is a bounded decreasing sequence, then $\{b_n\}$, where $b_n \equiv -a_n$ is a bounded increasing sequence. If $b_n \to b$, and if $a \equiv -b$, then $a_n \to a$, since $|a_n - a| = |-b_n + b| = |b_n - b|$.

VIII implies IX: VIII implies VII exactly as VII implies VIII. Hence VIII implies IX.

IX implies X: If $\{a_n\}$ is a bounded sequence, let $\{b_n\}$ be any monotonic subsequence of $\{a_n\}$, which exists by Theorem VII, § 1. Since $\{b_n\}$ is bounded and monotonic it converges by IX.

X implies the axiom of completeness: We first establish a lemma:

Lemma. *If X is true and if $p > 0$, then $p/n \to 0$ and $p/2^n \to 0$ as $n \to +\infty$.*

Proof of lemma. By X, since the sequence $\{p/n\}$ is bounded (by 0 and p), it contains a convergent subsequence. By Theorem X, § 1, any decreasing sequence that contains a convergent subsequence is itself convergent, and its limit is the infimum of the set of its values. The remaining details are given in the proof of the lemma to *II implies III.*

X implies the axiom of completeness, continued: Let C be any nonempty set, in the ordered field $\mathscr{G}$, that is bounded above, let A be the set of all points of $\mathscr{G}$ that are *not* upper bounds of C and let B be the set of all upper bounds of C. Then A and B are nonempty sets whose union is $\mathscr{G}$, and $x \in A$ and $y \in B$

imply $x < y$. Let a_1 be an arbitrary member of A and b_1 be an arbitrary member of B, and proceed inductively as in the proof of *II implies III* to define the increasing sequence $\{a_n\}$ of points of A and the decreasing sequence $\{b_n\}$ of points of B as follows: If $\frac{1}{2}(a_n + b_n) \in A$, call it a_{n+1} and define $b_{n+1} \equiv b_n$; if $\frac{1}{2}(a_n + b_n) \in B$, call it b_{n+1} and define $a_{n+1} \equiv a_n$. As shown in the proof of the preceding lemma, X implies the convergence of both $\{a_n\}$ and $\{b_n\}$, and as shown in the proof of *II implies III*, the limits of these sequences are the same: if $a_n \to a$ and $b_n \to b$, then $a = b$. We now wish to prove that $a = b$ is an upper bound of the set C, and proceed by assuming that there exists a point c of C greater than a: $c > a$. But since $a = \inf \{b_n\} < c$, there must be a term b_n such that $b_n < c$. This contradicts the fact that b_n is a member of the set B, and is therefore an upper bound of the set C. Finally, we wish to prove that $a = b$ is the *least* upper bound of C, and proceed by assuming that there exists a number u less than a such that u is an upper bound of C. But since $a = \sup \{a_n\} > u$, there must be a term a_n such that $a_n > u$. This contradicts the fact that a_n is a member of the set A, and is therefore not an upper bound of C. This completes the proof of the theorem.

Corollary. *If $\mathscr{G}$ is an ordered field in which any one of the ten conditions of the preceding theorem is satisfied, then $\mathscr{G}$ is Archimedean (§ 903).*

Proof. Each condition implies completeness, and completeness implies the Archimedean property (Theorem I, § 903).

3. CONTINUITY AND COMPLETENESS

Definition. *Let $\mathscr{G}$ be an ordered field and let f be a function from $\mathscr{G}$ to $\mathscr{G}$. Then f is **continuous at a point** a of its domain of definition† if and only if corresponding to an arbitrary positive member ϵ of $\mathscr{G}$ there exists a positive member δ of $\mathscr{G}$ such that the inequality $|x - a| < \delta$ implies the inequality $|f(x) - f(a)| < \epsilon$ for points x of the domain of f. A function is **continuous on a set** within its domain if and only if it is continuous at every point of that set. A function is **continuous** if and only if it is continuous on its domain. A function is **bounded above**, **bounded below**, or **bounded** if and only if its range of values is bounded above, bounded below, or bounded, respectively. A value $f(c)$ of the function f is the **maximum value** of f if and only if the inequality $f(c) \geq f(x)$ holds for every x in the domain of f. A value $f(d)$ of the function f is the **minimum value** of f if and only if the inequality $f(d) \leq f(x)$ holds for every x in the domain of f.*

---

† In § 1004, continuity for monotonic functions is defined in terms of suprema and infima. It is a straightforward exercise in any course in mathematical analysis at the level of advanced calculus to prove that the definition of § 1004 is subsumed under the present one (cf. *AC*, §§ 209 and 215, and Exs. 23 and 24, § 216).

Theorem. *Let $\mathscr{G}$ be an ordered field. Then each of the following is a necessary and sufficient condition for $\mathscr{G}$ to be complete†:*

I.   *Whenever f is a continuous function from $\mathscr{G}$ to $\mathscr{G}$ whose domain is a closed interval I, then the absolute value of f is bounded by a natural number, that is, there exists a natural number N such that $x \in I$ implies $|f(x)| \leq N$.*

II.   *Whenever f is a continuous function from $\mathscr{G}$ to $\mathscr{G}$ whose domain is a closed interval I, then f has a maximum value on I, that is, there exists a point c of I such that $f(c)$ is the maximum value of f.*

III.   *Whenever f is a continuous function from $\mathscr{G}$ to $\mathscr{G}$ whose domain is a closed interval I, then f has a minimum value on I, that is, there exists a point d of I such that $f(d)$ is the minimum value of f.*

IV.   *Whenever f is a continuous function from $\mathscr{G}$ to $\mathscr{G}$ whose domain is a closed interval I, then f has the* **intermediate value property** *on I; that is, if $f(u) < h < f(v)$, where $u \in I$ and $v \in I$, then there exists a point w of I such that $f(w) = h$.*

Proof.   That each of these conditions is necessary is a standard theorem in the elementary theory of real variables. For detailed proofs, see *AC*, § 218. We proceed now to establish sufficiency for each of the four conditions of the theorem. To this end we first prove two lemmas.

Lemma 1.   *In an ordered field $\mathscr{G}$ any linear function is everywhere continuous; that is, if a function f is defined by an equation of the form $f(x) = mx + c$, where m and c are arbitrary constants in $\mathscr{G}$, then f is continuous.*

Proof.   If $m = 0$, then the inequality $|f(x) - f(a)| < \epsilon$ reduces to $|c - c| = 0 < \epsilon$, which is trivially satisfied for all x, and the δ of the Definition can be taken equal to 1. If $m \neq 0$, then the inequality $|f(x) - f(a)| < \epsilon$ becomes $|(mx + c) - (ma + c)| = |m| \cdot |x - a| < \epsilon$, which is satisfied provided $\delta \equiv \epsilon/|m|$ and $|x - a| < \delta$.

Lemma 2.   *If $[a, b]$ is an arbitrary closed interval of $\mathscr{G}$ and if c and d are arbitrary members of $\mathscr{G}$, there exists a continuous function f defined on $[a, b]$ such that $f(a) = c$ and $f(b) = d$.*

Proof.   Let $f(x) \equiv c + \dfrac{d - c}{b - a}(x - a)$.

I implies completeness: Assume that I is true and that Property VII of the Theorem, § 2, fails, that is, there exists a bounded increasing sequence that does not converge. As in the proof of *II implies III* of the Theorem, § 2, there must exist a strictly increasing bounded sequence $\{a_n\}$ that does not converge. If $a \equiv a_1$, and if b is an upper bound of $\{a_n\}$ we shall define a continuous function f from $\mathscr{G}$ to $\mathscr{G}$ with domain $[a, b]$ whose absolute values are *not* bounded by any natural number N, as follows: If $a_n \leq x \leq a_{n+1}$, for

† Properties I, II, and III are due to K. W. T. Weierstrass (1815–1897, German), and property IV is due to B. Bolzano (1781–1848, of Prague, then Austrian).

$n = 1, 2, \cdots$, define $f(x)$ by "linear interpolation" with values $f(a_n) = n$ and $f(a_{n+1}) = n + 1$: $f(x) \equiv n + (x - a_n)/(a_{n+1} - a_n)$. If x is an upper bound of the set $\{a_n\}$ and $x \leq b$, define $f(x)$ to be 0. To see that f is continuous at every point x_0 of $[a, b]$ we examine the cases: (1) If x_0 is between a_n and a_{n+1}, f is continuous at x_0 by Lemma 2. (2) If x_0 is equal to some a_n, then f is continuous on each of the two intervals touching x_0 (one interval if $n = 1$). (3) If x_0 is an upper bound of $\{a_n\}$, let u be a smaller upper bound (which must exist since sup $\{a_n\}$ does not exist), and let $\delta \equiv x_0 - u$. Then $|x - x_0| < \delta$ implies $x > x_0 - \delta = u$ and hence $|f(x) - f(x_0)| = |0 - 0| < \epsilon$. We have thus obtained the desired contradiction to I.

II implies completeness: Assume that Property VII of the Theorem, § 2, fails and let $\{a_n\}$, a, and b be defined as in the proof of *I implies completeness.* Let the function f be defined by linear interpolation on $[a_n, a_{n+1}]$ so that $f(a_n) = (n - 1)/n$ for $n = 1, 2, \cdots$, by the formula $f(x) = (n - 1)/n + (x - a_n)/n(n + 1)(a_{n+1} - a_n)$, and let $f(x) \equiv 0$ if x is an upper bound of $\{a_n\}$ not exceeding b. Then (as before) f is continuous on $[a, b]$. Furthermore, since every value $f(x)$ is exceeded by some number of the form $(n - 1)/n$, f can have no maximum value on the interval $[a, b]$.

III implies completeness: Assume that Property VII of the Theorem, § 2, fails and let $\{a_n\}$, a, and b be defined as in the proof of *I implies completeness.* If f is the function just defined for II, then $g \equiv 1 - f$ ($g(x) \equiv 1 - f(x)$) is continuous on $[a, b]$ without having a minimum value there.

IV implies completeness: Assume that IV is true and that III of the Theorem, § 2, fails; that is, there exist nonempty sets A and B having no point in common, whose union is $\mathcal{G}$, and such that no point of A is a limit point of B and no point of B is a limit point of A. Let a be an arbitrary point of A and b be an arbitrary point of B, and assume for definiteness that $a < b$ (if $a > b$ change the notation). On the interval $[a, b]$ define the function f to be equal to 1 at all points of A and -1 at all points of B. Since this function fails to have the intermediate value property (nowhere on $[a, b]$ does it have the value 0), a contradiction to IV will have been obtained as soon as we show that f is continuous at every point x_0 of $[a, b]$. Assume for definiteness that $x_0 \in A$ (the same method applies if $x_0 \in B$), and let (u, v) be an open interval containing x_0 and no points of B. If $\delta \equiv \min(x_0 - u, v - x_0)$, then $\delta > 0$ and $|x - x_0| < \delta$ implies $u \leq x_0 - \delta < x < x_0 + \delta \leq v$ and hence $|f(x) - f(x_0)| = |1 - 1| = 0 < \epsilon$ for every point x in the domain of f. Therefore f is continuous, a contradiction to IV has been found, and the proof is complete.

4. CAUCHY SEQUENCES

As a preliminary to the next, and final, section we introduce a new concept, due to the French mathematician A. L. Cauchy (1789–1857), and prove a few of its properties.

Definition. *A sequence $\{a_n\}$ in an ordered field $\mathcal{G}$ is a **Cauchy sequence** if and only if corresponding to an arbitrary positive member ϵ of $\mathcal{G}$ there exists a natural number N such that the two inequalities $m > N$ and $n > N$ together imply $|a_m - a_n| < \epsilon$.*

Theorem I. *Every convergent sequence in an ordered field is a Cauchy sequence.*

Proof. Assume that $a_n \to a$ as $n \to +\infty$, and let $\epsilon > 0$ be given. Then corresponding to the positive element $\tfrac{1}{2}\epsilon$ there exists a natural number N such that $n > N$ implies $|a_n - a| < \tfrac{1}{2}\epsilon$. If simultaneously $m > N$ and $n > N$, then by the triangle inequality (VII, § 804), $|a_m - a_n| = |(a_m - a) + (-a_n + a)| \leqq |a_m - a| + |-a_n + a| = |a_m - a| + |a_n - a| < \tfrac{1}{2}\epsilon + \tfrac{1}{2}\epsilon = \epsilon$.

Theorem II. *Every Cauchy sequence in an ordered field is bounded.*

Proof. If $\epsilon = 1$, there exists a natural number N such that $n > N$ implies $|a_n - a_{N+1}| < 1$, or $a_{N+1} - 1 < a_n < a_{N+1} + 1$. Therefore, for every natural number n, min $\{a_1, a_2, \cdots, a_N, a_{N+1}\} - 1 \leqq a_n \leqq$ max $\{a_1, a_2, \cdots, a_N, a_{N+1}\} + 1$.

Theorem III. *Every Cauchy sequence, in an ordered field, that contains a convergent subsequence is convergent.*

Proof. Assume that $\{a_n\}$ is a Cauchy sequence, that $\{b_n\}$ is a subsequence with $b_n = a_{m_n}$, and that $b_n \to b$ as $n \to +\infty$, and let $\epsilon > 0$ be given. Let N_1 be a natural number such that $m > N_1$ and $n > N_1$ imply $|a_m - a_n| < \tfrac{1}{2}\epsilon$; let N_2 be a natural number such that $n > N_2$ implies $|b_n - b| < \tfrac{1}{2}\epsilon$, define $N \equiv$ max (N_1, N_2), and let $n > N$. Then $m_n \geqq n > N$ (cf. Example 4, § 1204), and $|a_n - b| = |(a_n - b_n) + (b_n - b)| \leqq |a_n - b_n| + |b_n - b| = |a_n - a_{m_n}| + |b_n - b| < \tfrac{1}{2}\epsilon + \tfrac{1}{2}\epsilon = \epsilon$.

5. THE ARCHIMEDEAN PROPERTY AND COMPLETENESS

In this final section we give three conditions each of which is both necessary and sufficient for completeness in an Archimedean ordered field (§ 903). If each of these is coupled with the assumption that the ordered field under consideration is Archimedean, then the present section completes a total list of eighteen alternative formulations of completeness (the axiom of completeness, the ten conditions of § 2, the four conditions of § 3, and the three new conditions about to be presented). (If Exercises 7–13, § 5, are included, the total number of statements of completeness is brought to twenty-five),

Theorem. *Let $\mathcal{G}$ be an Archimedean ordered field. Then each of the following is a necessary and sufficient condition for $\mathcal{G}$ to be complete:*

    I. *Every Cauchy sequence converges.*

II. *The **nested intervals property**† holds*: *If* $\{|a_n, b_n|\}$ *is a decreasing sequence of closed intervals, that is,* $[a_{n+1}, b_{n+1}] \subset [a_n, b_n]$ *for* $n = 1, 2, \cdots$, *then there exists at least one point* x *belonging to every interval of the sequence*: $x \in \underset{n}{\cap} [a_n, b_n]$.

III. *Whenever* f *is a continuous function from* $\mathscr{G}$ *to* $\mathscr{G}$ *whose domain is a closed interval* I, *then the absolute value of* f *is bounded, that is, there exists a positive member* p *of* $\mathscr{G}$ *such that* $x \in I$ *implies* $|f(x)| \leq p$.

Proof. *Necessity of I*: Assume $\mathscr{G}$ is complete, and let $\{a_n\}$ be a Cauchy sequence. By Theorem II, § 4, $\{a_n\}$ is bounded, and hence by Theorems VI and VII, § 1, $\{a_n\}$ contains a bounded monotonic subsequence $\{b_n\}$. By IX of the Theorem, § 2, $\{b_n\}$ converges and therefore, by Theorem III, § 4, $\{a_n\}$ converges.

Sufficiency of I: Assume that every Cauchy sequence in the Archimedean ordered field $\mathscr{G}$ converges and that $\mathscr{G}$ is not complete. By VII of the Theorem § 2, there must exist a bounded increasing sequence $\{a_n\}$ that does not converge. By assumption, then, $\{a_n\}$ is not a Cauchy sequence. This means that there exists a positive member ϵ of $\mathscr{G}$ such that for any natural number N there exist natural numbers m and n such that $m > N$, $n > N$, and $|a_m - a_n| \geq \epsilon$, and hence, if $m < n$, such that $a_n \geq a_m + \epsilon \geq a_N + \epsilon$. If $N = 1$, in particular, there exists a natural number n such that $a_n \geq a_1 + \epsilon$. Let m_1 be the least such natural number, so that $a_{m_1} \geq a_1 + \epsilon$. With a_{m_n} determined, let m_{n+1} be the least natural number k such that $k > m_n$ and $a_k \geq a_{m_n} + \epsilon$, so that $a_{m_{n+1}} \geq a_{m_n} + \epsilon$. In this way, by mathematical induction, a strictly increasing bounded subsequence $\{b_n\}$ of $\{a_n\}$ is defined, with $b_{n+1} \geq b_n + \epsilon$ for every natural number n, and $b_1 \geq a_1 + \epsilon$. Consequently (as is easily proved by mathematical induction), $b_n \geq a_1 + n\epsilon$ for every $n = 1, 2, \cdots$, and if b is an upper bound of the set $\{a_n\}$, we infer that $b \geq b_n \geq a_1 + n\epsilon$ for $n = 1, 2, \cdots$, and conclude that the inequality $n\epsilon \leq b - a_1$ holds for every natural number n. This contradicts the assumption that the ordered field $\mathscr{G}$ is Archimedean.

Necessity of II: This is a special case of IV of the Theorem, § 2.

Sufficiency of II: Assuming II, we wish to prove that the Archimedean ordered field $\mathscr{G}$ must be connected (cf. III of the Theorem, § 2). Accordingly, assume that $\mathscr{G}$ is an Archimedean ordered field that is not connected, and let $\mathscr{G}$ be represented as the union of two separated sets A and B. We proceed as in the proof of II implies III, of the Theorem, § 2, defining a bounded increasing sequence $\{a_n\}$ of points of A and a bounded decreasing sequence $\{b_n\}$ of points of B such that $a_n < b_n$ and $b_n - a_n = (b_1 - a_1)/2^{n-1}$ for every n. Since, in an Archimedean ordered field, if $p > 0$, $p/n < \epsilon$ whenever $n > p/\epsilon$, $p/n \to 0$ as $n \to +\infty$. Therefore, since $\{p/2^n\}$ is a subsequence of $\{p/n\}$, $p/2^n \to 0$ and hence $b_n - a_n \to 0$, as $n \to +\infty$. By the assumption

---

† Due to Georg Cantor (1845–1918, German). Also cf. IV of the Theorem, § 2, for the related *overlapping intervals property*.

II, since $\{[a_n, b_n]\}$ is a decreasing sequence of closed intervals, there exists an element c such that $a_n \leq c \leq b_n$ for all n. Since $0 \leq c - a_n \leq b_n - a_n$ and $0 \leq b_n - c \leq b_n - a_n$, $c - a_n \to 0$ and $b_n - c \to 0$ as $n \to +\infty$, or equivalently, $a_n \to c$ and $b_n \to c$ as $n \to +\infty$. Now, if $c \in A$, since $b_n \to c$, c is a limit point of B (contradiction, since A and B are separated), and if $c \in B$, since $a_n \to c$, c is a limit point of A (again, a contradiction).

Necessity of III: If $|f(x)|$ is bounded by a natural number N, then $|f(x)|$ is bounded by a positive member of $\mathscr{G}$ (namely, N).

Sufficiency of III: If $|f(x)|$ is bounded by a positive member p of $\mathscr{G}$, and if N is a natural number greater than p (which exists by the Archimedean property), then $|f(x)|$ is bounded by N.

6. EXERCISES

1. Let $\mathscr{Q}$ be the ordered field of rational numbers, which fails to be complete (cf. Theorem IV, § 905). Find a counter-example for each of the ten conditions of the Theorem, § 2. (For example, for I of that theorem find a nonempty set of rational numbers that is bounded below by a rational number but that has no greatest lower bound in $\mathscr{Q}$.)

2. Same as Exercise 1, for each of the four conditions of the Theorem, § 3.

3. Same as Exercise 1, for each of the three conditions of the Theorem, § 5.

★4. Let $\mathscr{H}$ be the non-Archimedean ordered field of rational functions from the real number system $\mathscr{R}$ to $\mathscr{R}$ defined and discussed in § 711–713. Find a counter-example for each of the ten conditions of the Theorem, § 2.

★5. Same as Exercise 4 for each of the four conditions of the Theorem, § 3.

6. Let $\mathscr{G}$ be an ordered field that is not complete and let p be an arbitrary positive member of $\mathscr{G}$. Prove that there exists a strictly increasing divergent sequence of positive members of $\mathscr{G}$ all of which are less than p.

In Exercises 7–10 let $\mathscr{G}$ be an ordered field. A function from $\mathscr{G}$ to $\mathscr{G}$ is said to be **one-to-one on its domain** D if and only if it is a one-to-one correspondence between the members of D and those of its range R (cf. § 310). A nonempty subset A of $\mathscr{G}$ is **compact** if and only if A has the Heine-Borel property; every open covering of A is reducible to a finite covering (cf. VI of the Theorem, § 2). Prove that each of the statements in Exercises 7–10, is a necessary and sufficient condition for $\mathscr{G}$ to be complete.

7. Whenever f is a continuous function from $\mathscr{G}$ to $\mathscr{G}$ whose domain is a closed interval and which is one-to-one on its domain, then the range of f is a closed interval. *Hint:* Use the example of the proof of II of the Theorem, § 3, for $a_1 \leq x \leq a_n$ for some n, and $f(x) \equiv b + 2 - x$ for $x > a_n$ for all n.

8. Whenever f is a continuous function from $\mathscr{G}$ to $\mathscr{G}$ whose domain is a closed interval and which is one-to-one on its domain, then the range of f is connected. (Cf. Ex. 7.)

9. Whenever f is a continuous function from $\mathscr{G}$ to $\mathscr{G}$ whose domain is a closed interval and which is one-to-one on its domain, then the range of f is compact. (Cf. Exs. 7–8.)

10. Whenever f is a continuous function from $\mathscr{G}$ to $\mathscr{G}$ whose domain is a closed interval and which is one-to-one on its domain, then f is monotonic.

In Exercises 11–12, let $\mathscr{G}$ be an Archimedean ordered field. The **inverse** of a function f from $\mathscr{G}$ to $\mathscr{G}$ that is one-to-one on its domain is the function from $\mathscr{G}$ to $\mathscr{G}$ obtained by interchanging the domain and range of f. A function f from $\mathscr{G}$ to $\mathscr{G}$ is said to be **uniformly continuous** on its domain D if and only if corresponding to an arbitrary positive member ϵ of $\mathscr{G}$ there exists a positive member δ of $\mathscr{G}$ such that the inequality $|x_1 - x_2| < \delta$ implies the inequality $|f(x_1) - f(x_2)| < \epsilon$ for points x_1 and x_2 of D. Prove that each of the statements, in Exercises 11–12, is a necessary and sufficient condition for $\mathscr{G}$ to be complete.

11. Whenever f is a continuous function from $\mathscr{G}$ to $\mathscr{G}$ whose domain is a closed interval and which is one-to-one on its domain, then the inverse function of f is continuous on the range R of f. *Hint:* Adjust the example of the hint of Exercise 7 by means of the formula $f(x) \equiv b + 1 - x$.

12. Whenever f is a continuous function from $\mathscr{G}$ to $\mathscr{G}$ whose domain is a closed interval I, then f is uniformly continuous on I. *Hint:* Use the example of the proof of part I of the Theorem, § 3.

13. Let $\mathscr{G}$ be an ordered field. A function f from $\mathscr{G}$ to $\mathscr{G}$ that is continuous and one-to-one on its domain D and whose inverse is continuous on the range R of f is said to be a **topological mapping** of D onto R. Prove that if $\mathscr{G}$ is not complete, then there exists a topological mapping of the closed unit interval $[0, 1]$ onto itself that is not monotonic. Prove indeed, that $\mathscr{G}$ is complete if and only if such a mapping is impossible. *Hint:* Let $0 < c_1 < \cdots < c_n < \cdots < \frac{1}{2}$, where $\{c_n\}$ does not converge, define $f(x) \equiv x$ if $0 \le x \le c_n$ for some n, or if $1 - c_n \le x \le 1$ for some n, and let $f(x) \equiv 1 - x$ if $c_n < x < 1 - c_n$ for every n.

Answers

§ 318, page 45

1. 138,618.
2. $2n^2 + 4n$.
3. $3280\frac{1}{3}$.
4. $2^n - 1$.
13. $\frac{1}{2}n(6n^2 + 3n - 1)$.
14. $n^2(2n^2 - 1)$.
15. $\frac{1}{3}n(n + 1)(n + 2)$.
16. $\frac{1}{12}n(n + 1)(n + 2)(3n + 1)$.
17. $\frac{1}{12}n(n + 1)(n + 2)(3n + 5)$.
18. $\frac{1}{15}n(n + 1)(n + 2)(3n^2 + 6n + 1)$.

§ 406, page 54

5. $41 \cdot 47$.
6. Prime.
7. $7 \cdot 7 \cdot 11 \cdot 13 \cdot 13$.
8. $89 \cdot 97 \cdot 101$.
9. 2, 3, 5, 7, 11, 13, 17, 19, 23, 29, 31, 37, 41, 43, 47, 53, 59, 61, 67, 71, 73, 79, 83, 89, 97, 101, 103, 107, 109, 113.
10. 3, 7, 31, 211, 2301.
15. 227, 229.
16. $20 = 3 + 17 = 7 + 13$, $30 = 7 + 23 = 11 + 19 = 13 + 17$, $40 = 3 + 37 = 11 + 29 = 17 + 23$, $50 = 3 + 47 = 7 + 43 = 13 + 37 = 19 + 31$, $60 = 7 + 53 = 13 + 47 = 17 + 43 = 19 + 41 = 23 + 37 = 29 + 31$.
17. $n = 41m, n = 42(41m + 1)$.

§ 409, page 60

1. 4; 24.
2. 1; 840.
3. 6; 546.
4. 10; 126,000.
5. 1; 111,111.
6. 12; 30,240.
7. $m = 2, n = 1, r = 1, s = 1$.
8. $m = 11, n = 16, r = 19, s = 13$.
9. $m = 6, n = 11, r = 2, s = 1$.
10. $m = 221, n = 55, r = 1, s = 4$.
11. $m = 56, n = 505, r = 496, s = 55$.
12. $m = 35, n = 17, r = 18, s = 37$.
14. 24.

§ 609, page 80

1. $x = 9$.
2. $x = 56$.
3. $x = 3$.
4. $x = 20$.
5. 12.
6. 11.
7. 32.
8. 89.
9. $n = 6$.
10. $n = 13$.
11. $x = 3, y = 4$.
12. $x = 3, y = 1, z = 2$.
15. $x = 0, y = 2$; $x = 1, y = 3$; $x = 2, y = 4$; $x = 3, y = 0$; $x = 4, y = 1$.
16. $x = 0, y = 2, z = 0$; $x = 1, y = 2, z = 2$; $x = 2, y = 2, z = 4$; $x = 3, y = 2, z = 1$; $x = 4, y = 2, z = 3$.
17. $x = 4, x = 7$.
18. No solutions.
23. 3.
24. 3.
27. $x = 4, y = 0, 2, 4, 6, 8, 10$; $x = 10, y = 1, 3, 5, 7, 9, 11$.

§ 710, page 98

4. $q(x) = \frac{3}{4}, r(x) = -\frac{1}{4}x - \frac{1}{2}$.
5. $q(x) = \frac{1}{2}x^2 - \frac{1}{4}x + \frac{3}{4}, r(x) = -\frac{3}{8}x + \frac{7}{4}$.
13. GCD: $x^2 - x + 1$, LCM: $3x^6 + 5x^5 + 9x^4 + 2x^3 + 9x^2 + 5x + 3$.

14. GCD: $x^3 - 5x - 3$, LCM: $x^8 - 25x^4 - 15x^3 - 9x^2 - 75x - 45$.
15. $\phi(x) = \frac{4}{3}x - \frac{5}{3}$, $\psi(x) = -x + \frac{4}{3}$.
16. $\phi(x) = \frac{4}{31}(6x + 7)$, $\psi(x) = -\frac{1}{31}(12x^3 + 2x^2 + 4x + 3)$.

§ 806, page 116

3. $-1 < x < 5$.
4. $x \leq -5$ or $x \geq -1$.
5. $x > 2$.
6. $x < 3$.
7. No solutions.
8. $-\infty < x < +\infty$.
9. $-3 \leq x \leq -1$ or $1 \leq x \leq 3$.
10. No solutions.
11. $x < -5/3$ or $x > 5$.
12. $x < -1$.
13. $4 < x < 6$.
14. $2 < x < 4$ or $6 < x < 12$.
15. No solutions.
16. $-\infty < x < +\infty$.
17. $|x| > |a|$.
18. $|x| < |a|$ or $x = a$.
19. $|x| \leq 2$ or $|x| > 3$.
20. $2 < |x| < 3$.
21. If $|a| = |b|$, no solutions; if $|a| < |b|$, $\left| x - \dfrac{b - a}{2} \right| > \dfrac{|a + b|}{2}$;

    if $|a| > |b|$, $\left| x - \dfrac{b - a}{2} \right| < \dfrac{|a + b|}{2}$.
22. If $a = b$, no solutions; if $a < b$, $-|b| < x < 0$ or $x > |b|$; if $a > b$, $0 < x < |b|$ or $x < -|b|$.
23. $x > a + |b|$.

§ 1211, page 183

1. 1,100,100; 1210; 84.
2. 10,010,000; 2100; 100.
3. 11,111,111; 3333; 193.
4. 1,111,100,111; 33,213; $6e3$.
15. 0.43002.
16. 0.110010001.
17. 0.30202.
18. 0.88.
19. $\dfrac{97}{275}$.
20. $\dfrac{7}{54}$.
21. $\dfrac{60}{101}$.
22. $\dfrac{9}{13}$.
23. $0.0101\overline{01}$.
24. $0.1313\overline{13}$.
25. 0.2.
26. $0.3737\overline{37}$.
27. $0.011011\overline{011}$.
28. $0.102120\overline{102120}$.
29. $0.333\overline{3}$.
30. $0.5186t3518\overline{6t3}$.
31. $\dfrac{3}{7}$; $\dfrac{2}{13}$; $\dfrac{7}{215}$; $\dfrac{11}{999}$; $\dfrac{13}{1727}$.
32. For $q = 3, 6, 9, 12, 15, 18, 24, 30$, and 36, the period π is 1; for $q = 11, 22$, and 33, $\pi = 2$; for $q = 27$ and 37, $\pi = 3$; for $q = 7, 13, 14, 21, 26, 28, 35$, and 39, $\pi = 6$; for $q = 31$, $\pi = 15$; for $q = 17$ and 34, $\pi = 16$; for $q = 19$ and 38, $\pi = 18$; for $q = 23$, $\pi = 22$; for $q = 29$, $\pi = 28$.
33. 1.10201.
34. 1.2013.
35. 1.225.
36. $1.4e7$.

LIST OF SPECIAL SYMBOLS

$m \mid k$	m divides k	§ 401, p. 48
(a, b)	GCD of a and b	§ 408, p. 56
$[a, b]$	LCM of a and b	§ 408, p. 57
(a, b, c)	GCD of a, b, and c	§ 409, p. 61
$[a, b, c]$	LCM of a, b, and c	§ 409, p. 61
$\mathscr{I}$	System of integers	§ 501, p. 62
$\mathscr{Q}$	Rational number system	§ 504, p. 66
$\sim$	Equivalence relation	§ 603, p. 70
Π	Partition	§ 603, p. 71
$m \mid k$	m divides k	§ 604, p. 72
$a \equiv b \pmod{m}$	Congruence modulo m	§ 604, p. 72
$[n]$	Equivalence class	§ 605, p. 73
$\deg f$	Degree of f	§ 702, p. 86
$f \mid g$	f divides g	§ 703, p. 88
(f, g)	GCD of f and g	§ 709, p. 96
$[f, g]$	LCM of f and g	§ 709, p. 97
$f{:}g$	Ordered pair of polynomials	§ 711, p. 100
$[f{:}g]$	Rational function	§ 711, p. 102
$\mathscr{H}$	Rational function system	§ 712, p. 103
(a, b)	Open interval	§ 802, p. 111
$[a, b]$	Closed interval	§ 802, p. 111
$(a, b], [a, b)$	Half-open intervals	§ 802, p. 111
$+ \infty$	Plus infinity	§ 803, p. 111
$- \infty$	Minus infinity	§ 803, p. 111
$(a, + \infty), [a, + \infty)$	Infinite intervals	§ 803, p. 112
$(- \infty, a), (- \infty, a]$	Infinite intervals	§ 803, p. 112
$(- \infty, + \infty)$	Real number system	§ 803, p. 112
$\lvert x \rvert$	Absolute value	§ 804, p. 112
$\sup (A)$	Supremum	§ 902, p. 119
l.u.b. (A)	Least upper bound	§ 902, p. 119
$\inf (A)$	Infimum	§ 902, p. 121
g.l.b. (A)	Greatest lower bound	§ 902, p. 121
$\{ \cdots \mid \cdots \}$	Set-builder notation	§ 902, p. 127
$\lVert f \rVert$	Norm of a function	§ 908, p. 133
Φ	$\{ a + b\sqrt{5} \mid a \in \mathscr{I}, b \in \mathscr{I} \}$	§ 1009, p. 143
$N(x)$	Norm of $x \in \Phi$	§ 1009, p. 144
$.d_1 d_2 \cdots d_m$	Terminating decimal	§ 1203, p. 167
$\lim_{n \to +\infty} a_n = a$	Limit of an increasing sequence	§ 1204, p. 169
$a_n \to a$	Limit of an increasing sequence	§ 1204, p. 169
$\sum a_n = \sum_{n=1}^{+\infty} a_n$	Infinite series	§ 1205, p. 171
$.d_1 d_2 \cdots d_n \cdots$	Nonterminating decimal	§ 1207, p. 173
$\bigcup_\alpha A_a$	Union of sets	§ 1, p. 192
$\bigcap_\alpha A_a$	Intersection of sets	§ 1, p. 192
$\lim_{n \to +\infty} a_n = a$	Limit of a sequence	§ 1, p. 192
$a_n \to a$	Limit of a sequence	§ 1, p. 192

INDEX

++

(The numbers refer to pages.)

211

A CATALOG OF SELECTED
DOVER BOOKS
IN SCIENCE AND MATHEMATICS

Mathematics–Bestsellers

HANDBOOK OF MATHEMATICAL FUNCTIONS: with Formulas, Graphs, and Mathematical Tables, Edited by Milton Abramowitz and Irene A. Stegun. A classic resource for working with special functions, standard trig, and exponential logarithmic definitions and extensions, it features 29 sets of tables, some to as high as 20 places. 1046pp. 8 x 10 1/2. 0-486-61272-4

ABSTRACT AND CONCRETE CATEGORIES: The Joy of Cats, Jiri Adamek, Horst Herrlich, and George E. Strecker. This up-to-date introductory treatment employs category theory to explore the theory of structures. Its unique approach stresses concrete categories and presents a systematic view of factorization structures. Numerous examples. 1990 edition, updated 2004. 528pp. 6 1/8 x 9 1/4. 0-486-46934-4

MATHEMATICS: Its Content, Methods and Meaning, A. D. Aleksandrov, A. N. Kolmogorov, and M. A. Lavrent'ev. Major survey offers comprehensive, coherent discussions of analytic geometry, algebra, differential equations, calculus of variations, functions of a complex variable, prime numbers, linear and non-Euclidean geometry, topology, functional analysis, more. 1963 edition. 1120pp. 5 3/8 x 8 1/2. 0-486-40916-3

INTRODUCTION TO VECTORS AND TENSORS: Second Edition–Two Volumes Bound as One, Ray M. Bowen and C.-C. Wang. Convenient single-volume compilation of two texts offers both introduction and in-depth survey. Geared toward engineering and science students rather than mathematicians, it focuses on physics and engineering applications. 1976 edition. 560pp. 6 1/2 x 9 1/4. 0-486-46914-X

AN INTRODUCTION TO ORTHOGONAL POLYNOMIALS, Theodore S. Chihara. Concise introduction covers general elementary theory, including the representation theorem and distribution functions, continued fractions and chain sequences, the recurrence formula, special functions, and some specific systems. 1978 edition. 272pp. 5 3/8 x 8 1/2. 0-486-47929-3

ADVANCED MATHEMATICS FOR ENGINEERS AND SCIENTISTS, Paul DuChateau. This primary text and supplemental reference focuses on linear algebra, calculus, and ordinary differential equations. Additional topics include partial differential equations and approximation methods. Includes solved problems. 1992 edition. 400pp. 7 1/2 x 9 1/4. 0-486-47930-7

PARTIAL DIFFERENTIAL EQUATIONS FOR SCIENTISTS AND ENGINEERS, Stanley J. Farlow. Practical text shows how to formulate and solve partial differential equations. Coverage of diffusion-type problems, hyperbolic-type problems, elliptic-type problems, numerical and approximate methods. Solution guide available upon request. 1982 edition. 414pp. 6 1/8 x 9 1/4. 0-486-67620-X

VARIATIONAL PRINCIPLES AND FREE-BOUNDARY PROBLEMS, Avner Friedman. Advanced graduate-level text examines variational methods in partial differential equations and illustrates their applications to free-boundary problems. Features detailed statements of standard theory of elliptic and parabolic operators. 1982 edition. 720pp. 6 1/8 x 9 1/4. 0-486-47853-X

LINEAR ANALYSIS AND REPRESENTATION THEORY, Steven A. Gaal. Unified treatment covers topics from the theory of operators and operator algebras on Hilbert spaces; integration and representation theory for topological groups; and the theory of Lie algebras, Lie groups, and transform groups. 1973 edition. 704pp. 6 1/8 x 9 1/4. 0-486-47851-3

Browse over 9,000 books at www.doverpublications.com

A SURVEY OF INDUSTRIAL MATHEMATICS, Charles R. MacCluer. Students learn how to solve problems they'll encounter in their professional lives with this concise single-volume treatment. It employs MATLAB and other strategies to explore typical industrial problems. 2000 edition. 384pp. 5 3/8 x 8 1/2. 0-486-47702-9

NUMBER SYSTEMS AND THE FOUNDATIONS OF ANALYSIS, Elliott Mendelson. Geared toward undergraduate and beginning graduate students, this study explores natural numbers, integers, rational numbers, real numbers, and complex numbers. Numerous exercises and appendixes supplement the text. 1973 edition. 368pp. 5 3/8 x 8 1/2. 0-486-45792-3

A FIRST LOOK AT NUMERICAL FUNCTIONAL ANALYSIS, W. W. Sawyer. Text by renowned educator shows how problems in numerical analysis lead to concepts of functional analysis. Topics include Banach and Hilbert spaces, contraction mappings, convergence, differentiation and integration, and Euclidean space. 1978 edition. 208pp. 5 3/8 x 8 1/2. 0-486-47882-3

FRACTALS, CHAOS, POWER LAWS: Minutes from an Infinite Paradise, Manfred Schroeder. A fascinating exploration of the connections between chaos theory, physics, biology, and mathematics, this book abounds in award-winning computer graphics, optical illusions, and games that clarify memorable insights into self-similarity. 1992 edition. 448pp. 6 1/8 x 9 1/4. 0-486-47204-3

SET THEORY AND THE CONTINUUM PROBLEM, Raymond M. Smullyan and Melvin Fitting. A lucid, elegant, and complete survey of set theory, this three-part treatment explores axiomatic set theory, the consistency of the continuum hypothesis, and forcing and independence results. 1996 edition. 336pp. 6 x 9. 0-486-47484-4

DYNAMICAL SYSTEMS, Shlomo Sternberg. A pioneer in the field of dynamical systems discusses one-dimensional dynamics, differential equations, random walks, iterated function systems, symbolic dynamics, and Markov chains. Supplementary materials include PowerPoint slides and MATLAB exercises. 2010 edition. 272pp. 6 1/8 x 9 1/4. 0-486-47705-3

ORDINARY DIFFERENTIAL EQUATIONS, Morris Tenenbaum and Harry Pollard. Skillfully organized introductory text examines origin of differential equations, then defines basic terms and outlines general solution of a differential equation. Explores integrating factors; dilution and accretion problems; Laplace Transforms; Newton's Interpolation Formulas, more. 818pp. 5 3/8 x 8 1/2. 0-486-64940-7

MATROID THEORY, D. J. A. Welsh. Text by a noted expert describes standard examples and investigation results, using elementary proofs to develop basic matroid properties before advancing to a more sophisticated treatment. Includes numerous exercises. 1976 edition. 448pp. 5 3/8 x 8 1/2. 0-486-47439-9

THE CONCEPT OF A RIEMANN SURFACE, Hermann Weyl. This classic on the general history of functions combines function theory and geometry, forming the basis of the modern approach to analysis, geometry, and topology. 1955 edition. 208pp. 5 3/8 x 8 1/2. 0-486-47004-0

THE LAPLACE TRANSFORM, David Vernon Widder. This volume focuses on the Laplace and Stieltjes transforms, offering a highly theoretical treatment. Topics include fundamental formulas, the moment problem, monotonic functions, and Tauberian theorems. 1941 edition. 416pp. 5 3/8 x 8 1/2. 0-486-47755-X

Browse over 9,000 books at www.doverpublications.com

Mathematics-Logic and Problem Solving

PERPLEXING PUZZLES AND TANTALIZING TEASERS, Martin Gardner. Ninety-three riddles, mazes, illusions, tricky questions, word and picture puzzles, and other challenges offer hours of entertainment for youngsters. Filled with rib-tickling drawings. Solutions. 224pp. 5 3/8 x 8 1/2.　　　　　　　　　　0-486-25637-5

MY BEST MATHEMATICAL AND LOGIC PUZZLES, Martin Gardner. The noted expert selects 70 of his favorite "short" puzzles. Includes The Returning Explorer, The Mutilated Chessboard, Scrambled Box Tops, and dozens more. Complete solutions included. 96pp. 5 3/8 x 8 1/2.　　　　　　　　　　0-486-28152-3

THE LADY OR THE TIGER?: and Other Logic Puzzles, Raymond M. Smullyan. Created by a renowned puzzle master, these whimsically themed challenges involve paradoxes about probability, time, and change; metapuzzles; and self-referentiality. Nineteen chapters advance in difficulty from relatively simple to highly complex. 1982 edition. 240pp. 5 3/8 x 8 1/2.　　　　　　　　　　0-486-47027-X

SATAN, CANTOR AND INFINITY: Mind-Boggling Puzzles, Raymond M. Smullyan. A renowned mathematician tells stories of knights and knaves in an entertaining look at the logical precepts behind infinity, probability, time, and change. Requires a strong background in mathematics. Complete solutions. 288pp. 5 3/8 x 8 1/2.

0-486-47036-9

THE RED BOOK OF MATHEMATICAL PROBLEMS, Kenneth S. Williams and Kenneth Hardy. Handy compilation of 100 practice problems, hints and solutions indispensable for students preparing for the William Lowell Putnam and other mathematical competitions. Preface to the First Edition. Sources. 1988 edition. 192pp. 5 3/8 x 8 1/2.　　　　　　　　　　0-486-69415-1

KING ARTHUR IN SEARCH OF HIS DOG AND OTHER CURIOUS PUZZLES, Raymond M. Smullyan. This fanciful, original collection for readers of all ages features arithmetic puzzles, logic problems related to crime detection, and logic and arithmetic puzzles involving King Arthur and his Dogs of the Round Table. 160pp. 5 3/8 x 8 1/2.

0-486-47435-6

UNDECIDABLE THEORIES: Studies in Logic and the Foundation of Mathematics, Alfred Tarski in collaboration with Andrzej Mostowski and Raphael M. Robinson. This well-known book by the famed logician consists of three treatises: "A General Method in Proofs of Undecidability," "Undecidability and Essential Undecidability in Mathematics," and "Undecidability of the Elementary Theory of Groups." 1953 edition. 112pp. 5 3/8 x 8 1/2.　　　　　　　　　　0-486-47703-7

LOGIC FOR MATHEMATICIANS, J. Barkley Rosser. Examination of essential topics and theorems assumes no background in logic. "Undoubtedly a major addition to the literature of mathematical logic." – *Bulletin of the American Mathematical Society.* 1978 edition. 592pp. 6 1/8 x 9 1/4.　　　　　　　　　　0-486-46898-4

INTRODUCTION TO PROOF IN ABSTRACT MATHEMATICS, Andrew Wohlgemuth. This undergraduate text teaches students what constitutes an acceptable proof, and it develops their ability to do proofs of routine problems as well as those requiring creative insights. 1990 edition. 384pp. 6 1/2 x 9 1/4.　　0-486-47854-8

FIRST COURSE IN MATHEMATICAL LOGIC, Patrick Suppes and Shirley Hill. Rigorous introduction is simple enough in presentation and context for wide range of students. Symbolizing sentences; logical inference; truth and validity; truth tables; terms, predicates, universal quantifiers; universal specification and laws of identity; more. 288pp. 5 3/8 x 8 1/2.　　　　　　　　　　0-486-42259-3

Mathematics–Algebra and Calculus

VECTOR CALCULUS, Peter Baxandall and Hans Liebeck. This introductory text offers a rigorous, comprehensive treatment. Classical theorems of vector calculus are amply illustrated with figures, worked examples, physical applications, and exercises with hints and answers. 1986 edition. 560pp. 5 3/8 x 8 1/2.　　0-486-46620-5

ADVANCED CALCULUS: An Introduction to Classical Analysis, Louis Brand. A course in analysis that focuses on the functions of a real variable, this text introduces the basic concepts in their simplest setting and illustrates its teachings with numerous examples, theorems, and proofs. 1955 edition. 592pp. 5 3/8 x 8 1/2.　　0-486-44548-8

ADVANCED CALCULUS, Avner Friedman. Intended for students who have already completed a one-year course in elementary calculus, this two-part treatment advances from functions of one variable to those of several variables. Solutions. 1971 edition. 432pp. 5 3/8 x 8 1/2.　　0-486-45795-8

METHODS OF MATHEMATICS APPLIED TO CALCULUS, PROBABILITY, AND STATISTICS, Richard W. Hamming. This 4-part treatment begins with algebra and analytic geometry and proceeds to an exploration of the calculus of algebraic functions and transcendental functions and applications. 1985 edition. Includes 310 figures and 18 tables. 880pp. 6 1/2 x 9 1/4.　　0-486-43945-3

BASIC ALGEBRA I: Second Edition, Nathan Jacobson. A classic text and standard reference for a generation, this volume covers all undergraduate algebra topics, including groups, rings, modules, Galois theory, polynomials, linear algebra, and associative algebra. 1985 edition. 528pp. 6 1/8 x 9 1/4.　　0-486-47189-6

BASIC ALGEBRA II: Second Edition, Nathan Jacobson. This classic text and standard reference comprises all subjects of a first-year graduate-level course, including in-depth coverage of groups and polynomials and extensive use of categories and functors. 1989 edition. 704pp. 6 1/8 x 9 1/4.　　0-486-47187-X

CALCULUS: An Intuitive and Physical Approach (Second Edition), Morris Kline. Application-oriented introduction relates the subject as closely as possible to science with explorations of the derivative; differentiation and integration of the powers of x; theorems on differentiation, antidifferentiation; the chain rule; trigonometric functions; more. Examples. 1967 edition. 960pp. 6 1/2 x 9 1/4.　　0-486-40453-6

ABSTRACT ALGEBRA AND SOLUTION BY RADICALS, John E. Maxfield and Margaret W. Maxfield. Accessible advanced undergraduate-level text starts with groups, rings, fields, and polynomials and advances to Galois theory, radicals and roots of unity, and solution by radicals. Numerous examples, illustrations, exercises, appendixes. 1971 edition. 224pp. 6 1/8 x 9 1/4.　　0-486-47723-1

AN INTRODUCTION TO THE THEORY OF LINEAR SPACES, Georgi E. Shilov. Translated by Richard A. Silverman. Introductory treatment offers a clear exposition of algebra, geometry, and analysis as parts of an integrated whole rather than separate subjects. Numerous examples illustrate many different fields, and problems include hints or answers. 1961 edition. 320pp. 5 3/8 x 8 1/2.　　0-486-63070-6

LINEAR ALGEBRA, Georgi E. Shilov. Covers determinants, linear spaces, systems of linear equations, linear functions of a vector argument, coordinate transformations, the canonical form of the matrix of a linear operator, bilinear and quadratic forms, and more. 387pp. 5 3/8 x 8 1/2.　　0-486-63518-X

Browse over 9,000 books at www.doverpublications.com

Mathematics–Probability and Statistics

BASIC PROBABILITY THEORY, Robert B. Ash. This text emphasizes the probabilistic way of thinking, rather than measure-theoretic concepts. Geared toward advanced undergraduates and graduate students, it features solutions to some of the problems. 1970 edition. 352pp. 5 3/8 x 8 1/2.   0-486-46628-0

PRINCIPLES OF STATISTICS, M. G. Bulmer. Concise description of classical statistics, from basic dice probabilities to modern regression analysis. Equal stress on theory and applications. Moderate difficulty; only basic calculus required. Includes problems with answers. 252pp. 5 5/8 x 8 1/4.   0-486-63760-3

OUTLINE OF BASIC STATISTICS: Dictionary and Formulas, John E. Freund and Frank J. Williams. Handy guide includes a 70-page outline of essential statistical formulas covering grouped and ungrouped data, finite populations, probability, and more, plus over 1,000 clear, concise definitions of statistical terms. 1966 edition. 208pp. 5 3/8 x 8 1/2.   0-486-47769-X

GOOD THINKING: The Foundations of Probability and Its Applications, Irving J. Good. This in-depth treatment of probability theory by a famous British statistician explores Keynesian principles and surveys such topics as Bayesian rationality, corroboration, hypothesis testing, and mathematical tools for induction and simplicity. 1983 edition. 352pp. 5 3/8 x 8 1/2.   0-486-47438-0

INTRODUCTION TO PROBABILITY THEORY WITH CONTEMPORARY APPLICATIONS, Lester L. Helms. Extensive discussions and clear examples, written in plain language, expose students to the rules and methods of probability. Exercises foster problem-solving skills, and all problems feature step-by-step solutions. 1997 edition. 368pp. 6 1/2 x 9 1/4.   0-486-47418-6

CHANCE, LUCK, AND STATISTICS, Horace C. Levinson. In simple, non-technical language, this volume explores the fundamentals governing chance and applies them to sports, government, and business. "Clear and lively ... remarkably accurate." – *Scientific Monthly.* 384pp. 5 3/8 x 8 1/2.   0-486-41997-5

FIFTY CHALLENGING PROBLEMS IN PROBABILITY WITH SOLUTIONS, Frederick Mosteller. Remarkable puzzlers, graded in difficulty, illustrate elementary and advanced aspects of probability. These problems were selected for originality, general interest, or because they demonstrate valuable techniques. Also includes detailed solutions. 88pp. 5 3/8 x 8 1/2.   0-486-65355-2

EXPERIMENTAL STATISTICS, Mary Gibbons Natrella. A handbook for those seeking engineering information and quantitative data for designing, developing, constructing, and testing equipment. Covers the planning of experiments, the analyzing of extreme-value data; and more. 1966 edition. Index. Includes 52 figures and 76 tables. 560pp. 8 3/8 x 11.   0-486-43937-2

STOCHASTIC MODELING: Analysis and Simulation, Barry L. Nelson. Coherent introduction to techniques also offers a guide to the mathematical, numerical, and simulation tools of systems analysis. Includes formulation of models, analysis, and interpretation of results. 1995 edition. 336pp. 6 1/8 x 9 1/4.   0-486-47770-3

INTRODUCTION TO BIOSTATISTICS: Second Edition, Robert R. Sokal and F. James Rohlf. Suitable for undergraduates with a minimal background in mathematics, this introduction ranges from descriptive statistics to fundamental distributions and the testing of hypotheses. Includes numerous worked-out problems and examples. 1987 edition. 384pp. 6 1/8 x 9 1/4.   0-486-46961-1

Browse over 9,000 books at www.doverpublications.com

Mathematics–Geometry and Topology

PROBLEMS AND SOLUTIONS IN EUCLIDEAN GEOMETRY, M. N. Aref and William Wernick. Based on classical principles, this book is intended for a second course in Euclidean geometry and can be used as a refresher. More than 200 problems include hints and solutions. 1968 edition. 272pp. 5 3/8 x 8 1/2.   0-486-47720-7

TOPOLOGY OF 3-MANIFOLDS AND RELATED TOPICS, Edited by M. K. Fort, Jr. With a New Introduction by Daniel Silver. Summaries and full reports from a 1961 conference discuss decompositions and subsets of 3-space; n-manifolds; knot theory; the Poincaré conjecture; and periodic maps and isotopies. Familiarity with algebraic topology required. 1962 edition. 272pp. 6 1/8 x 9 1/4.   0-486-47753-3

POINT SET TOPOLOGY, Steven A. Gaal. Suitable for a complete course in topology, this text also functions as a self-contained treatment for independent study. Additional enrichment materials make it equally valuable as a reference. 1964 edition. 336pp. 5 3/8 x 8 1/2.   0-486-47222-1

INVITATION TO GEOMETRY, Z. A. Melzak. Intended for students of many different backgrounds with only a modest knowledge of mathematics, this text features self-contained chapters that can be adapted to several types of geometry courses. 1983 edition. 240pp. 5 3/8 x 8 1/2.   0-486-46626-4

TOPOLOGY AND GEOMETRY FOR PHYSICISTS, Charles Nash and Siddhartha Sen. Written by physicists for physics students, this text assumes no detailed background in topology or geometry. Topics include differential forms, homotopy, homology, cohomology, fiber bundles, connection and covariant derivatives, and Morse theory. 1983 edition. 320pp. 5 3/8 x 8 1/2.   0-486-47852-1

BEYOND GEOMETRY: Classic Papers from Riemann to Einstein, Edited with an Introduction and Notes by Peter Pesic. This is the only English-language collection of these 8 accessible essays. They trace seminal ideas about the foundations of geometry that led to Einstein's general theory of relativity. 224pp. 6 1/8 x 9 1/4. 0-486-45350-2

GEOMETRY FROM EUCLID TO KNOTS, Saul Stahl. This text provides a historical perspective on plane geometry and covers non-neutral Euclidean geometry, circles and regular polygons, projective geometry, symmetries, inversions, informal topology, and more. Includes 1,000 practice problems. Solutions available. 2003 edition. 480pp. 6 1/8 x 9 1/4.   0-486-47459-3

TOPOLOGICAL VECTOR SPACES, DISTRIBUTIONS AND KERNELS, François Trèves. Extending beyond the boundaries of Hilbert and Banach space theory, this text focuses on key aspects of functional analysis, particularly in regard to solving partial differential equations. 1967 edition. 592pp. 5 3/8 x 8 1/2.
    0-486-45352-9

INTRODUCTION TO PROJECTIVE GEOMETRY, C. R. Wylie, Jr. This introductory volume offers strong reinforcement for its teachings, with detailed examples and numerous theorems, proofs, and exercises, plus complete answers to all odd-numbered end-of-chapter problems. 1970 edition. 576pp. 6 1/8 x 9 1/4.   0-486-46895-X

FOUNDATIONS OF GEOMETRY, C. R. Wylie, Jr. Geared toward students preparing to teach high school mathematics, this text explores the principles of Euclidean and non-Euclidean geometry and covers both generalities and specifics of the axiomatic method. 1964 edition. 352pp. 6 x 9.   0-486-47214-0

Browse over 9,000 books at www.doverpublications.com

Mathematics–History

THE WORKS OF ARCHIMEDES, Archimedes. Translated by Sir Thomas Heath. Complete works of ancient geometer feature such topics as the famous problems of the ratio of the areas of a cylinder and an inscribed sphere; the properties of conoids, spheroids, and spirals; more. 326pp. 5 3/8 x 8 1/2.   0-486-42084-1

THE HISTORICAL ROOTS OF ELEMENTARY MATHEMATICS, Lucas N. H. Bunt, Phillip S. Jones, and Jack D. Bedient. Exciting, hands-on approach to understanding fundamental underpinnings of modern arithmetic, algebra, geometry and number systems examines their origins in early Egyptian, Babylonian, and Greek sources. 336pp. 5 3/8 x 8 1/2.   0-486-25563-8

THE THIRTEEN BOOKS OF EUCLID'S ELEMENTS, Euclid. Contains complete English text of all 13 books of the Elements plus critical apparatus analyzing each definition, postulate, and proposition in great detail. Covers textual and linguistic matters; mathematical analyses of Euclid's ideas; classical, medieval, Renaissance and modern commentators; refutations, supports, extrapolations, reinterpretations and historical notes. 995 figures. Total of 1,425pp. All books 5 3/8 x 8 1/2.

Vol. I: 443pp.   0-486-60088-2
Vol. II: 464pp.   0-486-60089-0
Vol. III: 546pp.   0-486-60090-4

A HISTORY OF GREEK MATHEMATICS, Sir Thomas Heath. This authoritative two-volume set that covers the essentials of mathematics and features every landmark innovation and every important figure, including Euclid, Apollonius, and others. 5 3/8 x 8 1/2.

Vol. I: 461pp.   0-486-24073-8
Vol. II: 597pp.   0-486-24074-6

A MANUAL OF GREEK MATHEMATICS, Sir Thomas L. Heath. This concise but thorough history encompasses the enduring contributions of the ancient Greek mathematicians whose works form the basis of most modern mathematics. Discusses Pythagorean arithmetic, Plato, Euclid, more. 1931 edition. 576pp. 5 3/8 x 8 1/2.

0-486-43231-9

CHINESE MATHEMATICS IN THE THIRTEENTH CENTURY, Ulrich Libbrecht. An exploration of the 13th-century mathematician Ch'in, this fascinating book combines what is known of the mathematician's life with a history of his only extant work, the Shu-shu chiu-chang. 1973 edition. 592pp. 5 3/8 x 8 1/2.

0-486-44619-0

PHILOSOPHY OF MATHEMATICS AND DEDUCTIVE STRUCTURE IN EUCLID'S ELEMENTS, Ian Mueller. This text provides an understanding of the classical Greek conception of mathematics as expressed in Euclid's Elements. It focuses on philosophical, foundational, and logical questions and features helpful appendixes. 400pp. 6 1/2 x 9 1/4.   0-486-45300-6

BEYOND GEOMETRY: Classic Papers from Riemann to Einstein, Edited with an Introduction and Notes by Peter Pesic. This is the only English-language collection of these 8 accessible essays. They trace seminal ideas about the foundations of geometry that led to Einstein's general theory of relativity. 224pp. 6 1/8 x 9 1/4.   0-486-45350-2

HISTORY OF MATHEMATICS, David E. Smith. Two-volume history – from Egyptian papyri and medieval maps to modern graphs and diagrams. Non-technical chronological survey with thousands of biographical notes, critical evaluations, and contemporary opinions on over 1,100 mathematicians. 5 3/8 x 8 1/2.

Vol. I: 618pp.   0-486-20429-4
Vol. II: 736pp.   0-486-20430-8

Physics

THEORETICAL NUCLEAR PHYSICS, John M. Blatt and Victor F. Weisskopf. An uncommonly clear and cogent investigation and correlation of key aspects of theoretical nuclear physics by leading experts: the nucleus, nuclear forces, nuclear spectroscopy, two-, three- and four-body problems, nuclear reactions, beta-decay and nuclear shell structure. 896pp. 5 3/8 x 8 1/2. 0-486-66827-4

QUANTUM THEORY, David Bohm. This advanced undergraduate-level text presents the quantum theory in terms of qualitative and imaginative concepts, followed by specific applications worked out in mathematical detail. 655pp. 5 3/8 x 8 1/2. 0-486-65969-0

ATOMIC PHYSICS AND HUMAN KNOWLEDGE, Niels Bohr. Articles and speeches by the Nobel Prize–winning physicist, dating from 1934 to 1958, offer philosophical explorations of the relevance of atomic physics to many areas of human endeavor. 1961 edition. 112pp. 5 3/8 x 8 1/2. 0-486-47928-5

COSMOLOGY, Hermann Bondi. A co-developer of the steady-state theory explores his conception of the expanding universe. This historic book was among the first to present cosmology as a separate branch of physics. 1961 edition. 192pp. 5 3/8 x 8 1/2. 0-486-47483-6

LECTURES ON QUANTUM MECHANICS, Paul A. M. Dirac. Four concise, brilliant lectures on mathematical methods in quantum mechanics from Nobel Prize–winning quantum pioneer build on idea of visualizing quantum theory through the use of classical mechanics. 96pp. 5 3/8 x 8 1/2. 0-486-41713-1

THE PRINCIPLE OF RELATIVITY, Albert Einstein and Frances A. Davis. Eleven papers that forged the general and special theories of relativity include seven papers by Einstein, two by Lorentz, and one each by Minkowski and Weyl. 1923 edition. 240pp. 5 3/8 x 8 1/2. 0-486-60081-5

PHYSICS OF WAVES, William C. Elmore and Mark A. Heald. Ideal as a classroom text or for individual study, this unique one-volume overview of classical wave theory covers wave phenomena of acoustics, optics, electromagnetic radiations, and more. 477pp. 5 3/8 x 8 1/2. 0-486-64926-1

THERMODYNAMICS, Enrico Fermi. In this classic of modern science, the Nobel Laureate presents a clear treatment of systems, the First and Second Laws of Thermodynamics, entropy, thermodynamic potentials, and much more. Calculus required. 160pp. 5 3/8 x 8 1/2. 0-486-60361-X

QUANTUM THEORY OF MANY-PARTICLE SYSTEMS, Alexander L. Fetter and John Dirk Walecka. Self-contained treatment of nonrelativistic many-particle systems discusses both formalism and applications in terms of ground-state (zero-temperature) formalism, finite-temperature formalism, canonical transformations, and applications to physical systems. 1971 edition. 640pp. 5 3/8 x 8 1/2. 0-486-42827-3

QUANTUM MECHANICS AND PATH INTEGRALS: Emended Edition, Richard P. Feynman and Albert R. Hibbs. Emended by Daniel F. Styer. The Nobel Prize–winning physicist presents unique insights into his theory and its applications. Feynman starts with fundamentals and advances to the perturbation method, quantum electrodynamics, and statistical mechanics. 1965 edition, emended in 2005. 384pp. 6 1/8 x 9 1/4. 0-486-47722-3

Browse over 9,000 books at www.doverpublications.com

Physics

INTRODUCTION TO MODERN OPTICS, Grant R. Fowles. A complete basic undergraduate course in modern optics for students in physics, technology, and engineering. The first half deals with classical physical optics; the second, quantum nature of light. Solutions. 336pp. 5 3/8 x 8 1/2.   0-486-65957-7

THE QUANTUM THEORY OF RADIATION: Third Edition, W. Heitler. The first comprehensive treatment of quantum physics in any language, this classic introduction to basic theory remains highly recommended and widely used, both as a text and as a reference. 1954 edition. 464pp. 5 3/8 x 8 1/2.   0-486-64558-4

QUANTUM FIELD THEORY, Claude Itzykson and Jean-Bernard Zuber. This comprehensive text begins with the standard quantization of electrodynamics and perturbative renormalization, advancing to functional methods, relativistic bound states, broken symmetries, nonabelian gauge fields, and asymptotic behavior. 1980 edition. 752pp. 6 1/2 x 9 1/4.   0-486-44568-2

FOUNDATIONS OF POTENTIAL THERY, Oliver D. Kellogg. Introduction to fundamentals of potential functions covers the force of gravity, fields of force, potentials, harmonic functions, electric images and Green's function, sequences of harmonic functions, fundamental existence theorems, and much more. 400pp. 5 3/8 x 8 1/2.
0-486-60144-7

FUNDAMENTALS OF MATHEMATICAL PHYSICS, Edgar A. Kraut. Indispensable for students of modern physics, this text provides the necessary background in mathematics to study the concepts of electromagnetic theory and quantum mechanics. 1967 edition. 480pp. 6 1/2 x 9 1/4.   0-486-45809-1

GEOMETRY AND LIGHT: The Science of Invisibility, Ulf Leonhardt and Thomas Philbin. Suitable for advanced undergraduate and graduate students of engineering, physics, and mathematics and scientific researchers of all types, this is the first authoritative text on invisibility and the science behind it. More than 100 full-color illustrations, plus exercises with solutions. 2010 edition. 288pp. 7 x 9 1/4.   0-486-47693-6

QUANTUM MECHANICS: New Approaches to Selected Topics, Harry J. Lipkin. Acclaimed as "excellent" (*Nature*) and "very original and refreshing" (*Physics Today*), these studies examine the Mössbauer effect, many-body quantum mechanics, scattering theory, Feynman diagrams, and relativistic quantum mechanics. 1973 edition. 480pp. 5 3/8 x 8 1/2.   0-486-45893-8

THEORY OF HEAT, James Clerk Maxwell. This classic sets forth the fundamentals of thermodynamics and kinetic theory simply enough to be understood by beginners, yet with enough subtlety to appeal to more advanced readers, too. 352pp. 5 3/8 x 8 1/2.   0-486-41735-2

QUANTUM MECHANICS, Albert Messiah. Subjects include formalism and its interpretation, analysis of simple systems, symmetries and invariance, methods of approximation, elements of relativistic quantum mechanics, much more. "Strongly recommended." – *American Journal of Physics*. 1152pp. 5 3/8 x 8 1/2.   0-486-40924-4

RELATIVISTIC QUANTUM FIELDS, Charles Nash. This graduate-level text contains techniques for performing calculations in quantum field theory. It focuses chiefly on the dimensional method and the renormalization group methods. Additional topics include functional integration and differentiation. 1978 edition. 240pp. 5 3/8 x 8 1/2.
0-486-47752-5

Browse over 9,000 books at www.doverpublications.com

Physics

MATHEMATICAL TOOLS FOR PHYSICS, James Nearing. Encouraging students' development of intuition, this original work begins with a review of basic mathematics and advances to infinite series, complex algebra, differential equations, Fourier series, and more. 2010 edition. 496pp. 6 1/8 x 9 1/4. 0-486-48212-X

TREATISE ON THERMODYNAMICS, Max Planck. Great classic, still one of the best introductions to thermodynamics. Fundamentals, first and second principles of thermodynamics, applications to special states of equilibrium, more. Numerous worked examples. 1917 edition. 297pp. 5 3/8 x 8. 0-486-66371-X

AN INTRODUCTION TO RELATIVISTIC QUANTUM FIELD THEORY, Silvan S. Schweber. Complete, systematic, and self-contained, this text introduces modern quantum field theory. "Combines thorough knowledge with a high degree of didactic ability and a delightful style." – *Mathematical Reviews.* 1961 edition. 928pp. 5 3/8 x 8 1/2. 0-486-44228-4

THE ELECTROMAGNETIC FIELD, Albert Shadowitz. Comprehensive undergraduate text covers basics of electric and magnetic fields, building up to electromagnetic theory. Related topics include relativity theory. Over 900 problems, some with solutions. 1975 edition. 768pp. 5 5/8 x 8 1/4. 0-486-65660-8

THE PRINCIPLES OF STATISTICAL MECHANICS, Richard C. Tolman. Definitive treatise offers a concise exposition of classical statistical mechanics and a thorough elucidation of quantum statistical mechanics, plus applications of statistical mechanics to thermodynamic behavior. 1930 edition. 704pp. 5 5/8 x 8 1/4. 0-486-63896-0

INTRODUCTION TO THE PHYSICS OF FLUIDS AND SOLIDS, James S. Trefil. This interesting, informative survey by a well-known science author ranges from classical physics and geophysical topics, from the rings of Saturn and the rotation of the galaxy to underground nuclear tests. 1975 edition. 320pp. 5 3/8 x 8 1/2. 0-486-47437-2

STATISTICAL PHYSICS, Gregory H. Wannier. Classic text combines thermodynamics, statistical mechanics, and kinetic theory in one unified presentation. Topics include equilibrium statistics of special systems, kinetic theory, transport coefficients, and fluctuations. Problems with solutions. 1966 edition. 532pp. 5 3/8 x 8 1/2. 0-486-65401-X

SPACE, TIME, MATTER, Hermann Weyl. Excellent introduction probes deeply into Euclidean space, Riemann's space, Einstein's general relativity, gravitational waves and energy, and laws of conservation. "A classic of physics." – *British Journal for Philosophy and Science.* 330pp. 5 3/8 x 8 1/2. 0-486-60267-2

RANDOM VIBRATIONS: Theory and Practice, Paul H. Wirsching, Thomas L. Paez and Keith Ortiz. Comprehensive text and reference covers topics in probability, statistics, and random processes, plus methods for analyzing and controlling random vibrations. Suitable for graduate students and mechanical, structural, and aerospace engineers. 1995 edition. 464pp. 5 3/8 x 8 1/2. 0-486-45015-5

PHYSICS OF SHOCK WAVES AND HIGH-TEMPERATURE HYDRO DYNAMIC PHENOMENA, Ya B. Zel'dovich and Yu P. Raizer. Physical, chemical processes in gases at high temperatures are focus of outstanding text, which combines material from gas dynamics, shock-wave theory, thermodynamics and statistical physics, other fields. 284 illustrations. 1966–1967 edition. 944pp. 6 1/8 x 9 1/4. 0-486-42002-7

Browse over 9,000 books at www.doverpublications.com